MAJOR DEPRESSIVE DISORDER - COGNITIVE AND NEUROBIOLOGICAL MECHANISMS

Edited by **Yong-Ku Kim**

Major Depressive Disorder - Cognitive and Neurobiological Mechanisms
http://dx.doi.org/10.5772/58711
Edited by Yong-Ku Kim

Contributors

Xiaobin Zhang, Jin Li, Sha Weiwei, Ru Bu, Hwa-Young Lee, Massimo Cocchi, Lucio Tonello, Fabio Gabrielli, Zhengzhi Feng, Xiao Liu, Xiaoxia Wang, Lifei Wang, Keyu Liu, Xiao Chen, Qin Dai, Sang-Won Jeon, Yong-Ku Kim, Tabitha Payne, Madeleine Thompson

Published by InTech

Janeza Trdine 9, 51000 Rijeka, Croatia

Publishing Process Manager Iva Lipovic

Technical Editor InTech DTP team

Cover InTech Design team

First published June, 2015

Printed in Croatia

Additional hard copies can be obtained from orders@intechopen.com

Major Depressive Disorder - Cognitive and Neurobiological Mechanisms, Edited by Yong-Ku Kim

p. cm.

ISBN 978-953-51-2129-9

For EU product safety concerns:
IN TECH d.o.o., Prolaz Marije Krucifikse Kozulić 3, 51000 Rijeka, Croatia,
info@intechopen.com or visit our website at intechopen.com.

Meet the editor

Professor Yong-Ku Kim graduated in Medicine from the Korea University, College of Medicine, Seoul, South Korea in 1987. He received his master degree in 1991 and PhD in Psychiatry in 1998, both from Department of Psychiatry, Korea University. His professional activities extend beyond the area of pure clinical or research work. He has well over 170 publications in peer-reviewed international journals and twenty book chapters in the area of biological psychiatry. He is currently on the editorial boards of some leading academic journals in the field of neurobiology, including Progress in Neuropsychopharmacology and Biological Psychiatry, Psychiatry Investigation, World Journal Psychiatry, Annals of Depression and Anxiety, and Psychiatry Journal. He has mainly conducted research in psychoimmunology, biological marker of suicide, genetic polymorphism in biological psychiatry.

Contents

Preface

Major depressive disorder (MDD) is a common disorder that is widely distributed in the population and is usually associated with substantial symptom severity and role impairment. According to the fifth edition of DSM (DSM-5), major depressive episodes are characterized by depressed mood or loss of interest, or both, for more than 2 weeks, and four symptoms from a list that includes changes in appetite and weight, changes in sleep and psychomotor activity, lack of energy, feelings of guilt, problems thinking and making decisions and recurrent thoughts of death or suicide. A diagnosis of MDD can be made if a person suffers at least 1 such episode without ever experiencing mania or hypomania. However, most people with MDD have multiple episodes.

MDD is a complex and heterogeneous disorder, both on a phenotypical and on a biological level. MDD is not likely to result from a single gene or a single external event, but may be caused by a complex interaction between genes and the environment in susceptible persons. Until now, a gene or series of genes that cause MDD have not been identified. Rather, genes clearly can be a risk factor for developing depression, increasing the likelihood that severe environmental stress will precipitate the onset of this disease. Thus, a combination of genetics, early life stress, and ongoing stress may ultimately determine individual responsiveness to stress and vulnerability to MDD.

Gene-environment interactions in the pathophysiology of MDD leads to advancement in personalized medicine by means of genotyping for inter-individual variability in drug action and metabolism. Gene-environment interactions may explain why some subjects become depressed while others remain unaffected.

The aim of this book is to describe current knowledge of MDD from the point of view of neurobiology, molecular genetics and cognition. In this book, Lee et al. comprehensively review the neurobiological mechanisms involved in the pathogenesis of depressive subtypes, especially in melancholic and atypical depression. An understanding of the pathogenesis of MDD based on subtypes is helpful in predicting an individual's response to treatment for MDD. For example, melancholic depression is associated with hyperactivity of the hypothalamic-pituitary-adrenal (HPA) axis whereas atypical depression is associated with hypoactivity of the HPA axis.

A number of studies have demonstrated that stress and depression can decrease brain-derived neurotrophic factor (BDNF) expression. BDNF shows an antidepressant-like effect by antagonizing learned helplessness in stressed animals. Zhang et al. discussed the critical role of BDNF in the pathogenesis and pathophysiology of MDD. Such studies have led to the formulation of the neurotrophic hypothesis of depression, which proposes that reduced neurotrophic factor levels predispose patients to depression, whereas increases in neurotrophic factors produce an antidepressant action.

The pathogenesis of MDD remains unclear. MDD results from multiple genes interacting with environmental factors such as early stressful life events.

Cocchi et al. addressed the molecular and quantum approach to psychopathology and consciousness from theory to practice. On the basis of their hypothesis, MDD is a real disease with specific molecular features and expressions of consciousness, according to the concept of symmetry breaking.

Jeon et al. discussed how neuroimaging research in MDD can reveal the relationship between psychotherapy and brain function through a systemic and critical review of longitudinal studies published to date. The mechanisms of psychotherapy and medication appear divergent. To explain these different mechanisms of action, it has been suggested that, while psychotherapy may exert top-down effects targeting mainly frontal cortical regions and reducing dysfunctional thought processes, pharmacotherapy may produce bottom-up changes by disengaging the ventral and limbic regions mediating attention to personally relevant emotional and environmental stimuli.

Feng et al. addressed the neurobiological mechanism of negative cognitive bias in MDD. Beck's cognitive-neurobiological model suggests that cognitive bias in MDD is due to maladaptive bottom-up processes generally perpetuated by attenuated cognitive control, and has provided an evidence-based framework for conceptualizing and treating MDD.

Payne et al. examined the potential relationship between mental processing speed and depression in college students and confirmed that individuals reporting symptoms of moderate to severe depression have impaired mental speed, as measured by the discrimination version of the inspection time measure. The findings of that study are consistent with the cognitive slowing hypothesis, in which mental speed is disrupted and may lead to psychomotor slowing.

I would like to thank all the contributors for their valuable time spent preparing manuscripts and I believe this book will function as a step on the path toward the ultimate goal of predicting, preventing, and treating MDD.

Yong-Ku Kim, M.D, Ph.D
Professor
Department of Psychiatry
College of Medicine
Korea University
Republic of Korea

Chapter 1

Different Mechanisms Between Melancholic and Atypical Depression

Hwa-Young Lee and Yong-Ku Kim

Additional information is available at the end of the chapter

http://dx.doi.org/10.5772/59959

1. Introduction

Major depressive disorder (MDD) is very prevalent and disable psychiatric disorder with prevalence estimates ranging 5% to 20% [1, 2] and has been a growing public health concern due to its recurrent and lethal nature. According to projections, MDD will become the second leading cause of disability worldwide by the year 2020.[3]

Major depressive disorder is considered to be a clinically heterogeneous disorder and the diagnosis is based on a patient's symptoms, not on any laboratory tests. So, the pathophysiology of MDD is not clear. MDD results from multiple genetic factors interacting with many various environmental factors, such as childhood adversity and many life stressful events.[4]

Although work in this area has been inconclusive, many animal, post-mortem, clinical, and genetic studies have produced results implicating at least three neurobiological systems in the pathogenesis of MDD: the monoamine system, the hypothalamic-pituitary-adrenal axis (HPA axis), and neuroplasticity. Additionally, other biological factors, including inflammatory markers, neurophysiologic markers, and neuroimaging markers may be associated with MDD.

Although recent decades have witnessed a tremendous revolution in the development of antidepressant drugs, the neurochemical effects that underlie the therapeutic actions of these agents remain largely unknown.

There has been increasing data showing that depressive disorders are heterogeneous, and therefore, can vary with regard to HPA axis activity, immune function, and treatment response. Considering the biological mechanisms of depressive subtypes, it is helpful to understand the pathogenesis in order to more accurately predict an individual's response to a specific treatment for depression.

Melancholic depression is distinguished by a loss of appetite and sleep; melancholic patients are usually anxious and lose responsiveness to their environments. Those with melancholic depression tend to feel worse in the morning, while those with atypical depression generally feel worse in the evening.

Atypical depression is a subtype of depression that the DSM-5 defines as having the characteristics of reactive mood (including the ability to respond emotionally to environmental cues), increased appetite, hypersomnia, leaden paralysis, and interpersonal rejection sensitivity.[5] Patients with atypical depressive episodes generally have a younger mean age of onset than those with typical depression.[3-7] Individuals with atypical depression are 2- to 3-fold more likely to be women and often have a more chronic, unrelenting course of depression than individuals with typical depression. In a sample of 8116 individuals aged 15–64 years, 17.1% of the patients with a diagnosis of MDD had a history of atypical depression.[13] Of 1500 outpatients studied in the STAR*D (Sequenced Treatment Alternatives to Relieve Depression) trial, 18.1% of patients in the trial had depression with atypical features, and women were found to be 70% more likely to have atypical depression.[5] Studies of clinical populations have shown 18–36% of patients with MDD present with atypical depression.[6]

In this chapter, we discuss the biological mechanisms involved in the pathogenesis of depressive subtypes.

2. Different mechanisms between melancholic and atypical depression

2.1. Hormonal axis

MDD generally features the hyperactivity of the hypothalamic-pituitaryadrenal (HPA), a neuroendocrine abnormality[7] In particular, the majority of depressed patients exhibit hypersecretion of cortisol in their plasma, urine, and cerebrospinal fluid (CSF), and a hyperactive cortisol response to adrenocorticotrophic hormone (ACTH).[8-10] To explain the pathophysiology of MDD, the corticosteroid receptor hypothesis has been proposed. It focuses on corticosteroid receptor resistance, which results in a reduction of the negative feedback of cortisol, an increased production of corticotropin-releasing hormone (CRH), and ultimately, hypercortisolism. [8]

Interestingly, the serotonin (5-HT) system is affected by both cortisol and CRH. [9, 11] 5-HT transmission is stimulated by glucocorticoids(GCs) during the stress response.[12] Conversely, during chronic psychosocial stress, 5-HT transmission is impaired and noradrenergic transmission in the hippocampus is suppressed resulting in hypercortisolism, which is similar to the state of depression.[13] Depression may have a genetic component: it has been reported that HPA axis dysregulation could be a genetically determined trait that contributes to an increased susceptibility for depression. However, since the trait is found in both affected subjects and in healthy relatives with a high familial risk, the HPA axis is an interesting candidate endophenotype for affective disorders. [14, 15] Studies regarding the causes of the dysregulated HPA axis in depression have mainly focused on two elements: i) glucocorticoid receptor (GR) feedback mechanisms and ii) the CRH signaling system.

A reduced sensitivity to cortisol, leading to an impaired negative feedback mechanism has been attributed to resistant GR function. [16] In contrast, the CRH peptide mediates the regulation of the HPA axis as well as autonomic and behavioral responses during stress.[17] Furthermore, the functional action of antidepressants has been linked to the HPA axis.[8, 18] Consequently, a proper clinical response to antidepressant treatment includes the normalization of the dysregulated HPA axis..[9, 19]

Susceptibility to MDD has also been associated with Bcl1 polymorphism, and it was found to be predictive of treatment response.[20] Genetic association studies have yielded preliminary evidence for a role of GR genetic variations in the genetic vulnerability for MDD. Pharmacogenetic studies have investigated polymorphisms in components of the HPA axis in relation to the treatment response to antidepressants. In line with a SNP in the CRH-binding protein[21], a SNP in the FKPB5 protein involved in the regulation of GR sensitivity has been reported to be associated with response to citalopram (Lekman et al., 2008). SNPs in the FKPB5 protein were also associated with response to citalopram in a large cohort study in Munich to TCAs and SSRIs.[22] Taken together, the evidence for a role of GR and the GR gene in the neurobiology of MDD is building rapidly.[23]

Most studies in melancholic depression have found that relative HPA axis hyperactivity occurs, compared to non-depressed states, and that this is more likely to occur in more severe forms of depression.[24] In addition to increased corticotropin-releasing hormone (CRH) production, the overdrive in the HPA axis in depression has been attributed to both glucocorticoids feedback insensitivity and to the overproduction of other corticotrophin secretagogues insensitive to glucocorticoid feedback, such as arginine vasopressin.[25, 26] CRH and arginine vasopressin (AVP) are the main secretagogues of the HPA/stress system. Produced in the parvicellular division of the hypothalamic paraventricular nucleus, the release of these peptides is influenced by input from monoaminergic neurones. In depression, anterior pituitary CRH1 receptors are down-regulated and the resultant response to CRH infusion is blunted. By contrast, vasopressin V3 receptors on the anterior pituitary show an enhanced response to AVP stimulation and this enhancement plays a key role in maintaining HPA hyperactivity.[26]

Contrary to melancholic depression, atypical depression has reversed vegetative symptoms, i.e. hypersomnia and hyperphagia. The patients with melancholic depression show hypercortisolism and more disturbed sleep, as is strongly associated with high nocturnal ACTH and cortisol secretion.[27] Weight loss is correlated with hypercortisolism and dexamethasone non-suppression.[28, 29] Moreover, depressed patients without hypersomnia or increased appetite were shown to have elevated urinary cortisol concentrations as compared to normal morning plasma cortisol levels, as well as and a higher incidence of cortisol non-suppression after dexamethasone compared to normal subjects.[30] In contrast to typically depressed patients, those with hypersomnia and hyperphagia showed no change in morning plasma cortisol and DST.[30, 31]

It has been presented that a relatively hyperactive HPA axis leads to the symptoms of melancholic depression, while a relatively hypoactive stress response leads to the symptoms of atypical depression.[32] That is, CRH hypersecretion and hyposecretion correlate with the symptomatic pattern of melancholic and atypical depression, respectively. A recent meta-analysis of 40 years of HPA axis research conducted has identified a pattern of relative hypocortisolemia in atypical depression as compared to melancholic depression.[33]

Antonijevic expanded the concept and proposed that clinically relevant differences in the underlying pathophysiology in patients with depression exist, and that the identification of distinct endophenotypes for MDD will not only improve our understanding of the disease, but will also contribute to more specific treatment strategies.[34] Concerning pharmacological treatment, it was reported that the group of patients with atypical depression showed a significantly higher cortisol response to desipramine, a relatively selective noradrenaline reuptake inhibitor, than the group with no atypical symptoms and the group with mood reactivity as the only atypical symptom, indicating that atypical depression may be associated with a smaller impairment of the noradrenaline neurotransmitter system.[35] Similarly, hypersecretion of corticotropin-releasing hormone (CRH) and the resulting hypercortisolism were not found in patients with atypical depression.[36]

2.2. Neurotransmitter system

It has been hypothesized that a deficiency in serotonin is an essential determinant in the pathogenesis of MDD. Consequently, the serotonin system has been thoroughly investigated in a variety of MDD studies. The serotonin system projects from the dorsal raphe nucleus to all regions of the brain, including the cerebral cortex and hippocampus. In depressed patients, the diminished function and activity of the serotonin system has been confirmed in postmortem serotonin transporter and serotonin receptor studies. [citation?]

In suicide victims with MDD, enhanced radioligand binding of an agonist to inhibitory serotonin-1A autoreceptors in the human dorsal raphe nucleus was found, supporting the hypothesis regarding the reduced activity of serotonin neurons.[37] There appears to be a strong trend of decreased 5-HT1A receptor expression in MDD. Biochemically, the polymorphism of the C-1019G promoter (rs6295), a genetic variant of the 5-HT1A receptor, has shown to have the G allele is more frequently in MDD.[38]

Imipramine may be a putative biological marker of depressive disorder. It binds to the serotonin transporter (5-HTT) on platelets, and decreased imipramine binding may indicate depressive disorder. A meta-analysis showed a highly significant decrease in maximal binding values in depressed subject groups, which was further shown to be even greater among those who had been free of medication for 4 weeks at the time of investigation. [39]

Tryptophan hydroxylase (TPH), which has two isoforms (TPH1 and TPH2), is a of the rate-limiting factors in serotonin synthesis. Significantly higher numbers and densities of TPH immunoreactive neurons in the dorsal raphe nuclei of alcohol-dependent, depressed suicide victims compared to controls have been reported. [40] A deficient or impaired serotonin system seems to correlate with depressive disorders, as evidenced by studies on the serotonin receptor, TPH, and 5-HTT.

The norepinephrine (NE) system has been studied in depression, particularly the action of NE reuptake inhibitors. Monoaminergic neurobiology, including norepinephrine, has been associated with the mechanism of action of serotonin norepinephrine reuptake inhibitors (SNRIs), norepinephrine dopamine reuptake inhibitors (NDRIs), tricyclic and monoamine oxidase inhibitor antidepressants. Furthermore, mirtazapine's antidepressant effect seems to be due to the dual enhancement of central noradrenergic and serotonergic neurotransmission stemming from a blockade of adrenergic $\alpha 2$ receptors. [41-43]

The dopamine (DA) system has been reported to be highly associated with the symptomatology of depression, as the proposed pathogenesis of melancholic depression involves decreased DA transmission.[44]

In addition to HPA axis activity, distinct alterations of the serotonergic system may also play critical roles in the melancholic and atypical phenotypes, namely a reduced restraint of serotonin synthesis via 5-HT(1A) autoreceptors in the former, and primarily through reduced serotonin synthesis in the latter. Thus, the melancholic subtype with noradrenergic and HPA axis overdrive seems to be associated with reduced 5-HT1A autoreceptor function and, therefore, enhanced serotonergic activation of the HPA axis, as well as an acute phase immune reaction. The latter contributes to HPA axis stimulation and reduces the negative feedback inhibition from corticosteroid receptors. The resulting hypercortisolism can further impair 5-HT1A receptor functions, leading to a vicious circle, which may not be effectively resolved by most selective serotonin reuptake inhibitors (SSRIs).[32, 45] On the other hand, patients with AD and low HPA activity seem to have reduced noradrenergic and serotonergic afferent stimulation, possibly because of reduced serotonin (5-HT) synthesis and, unlike melancholic patients, an unimpaired 5-HT1A autoreceptor function.[32, 46] Moreover, MAOIs have been repeatedly found to be more effective for treating atypical depression than tricyclic antidepressants (TCAs), which have potent noradrenergic properties. This distinction between MAOIs and TCAs may indicate different biological mechanisms at work in patients with atypical or melancholic depression.[18,19] In fact, some researchers have suggested that serotonergic neurotransmission is more relevant than noradrenergic transmission to the pathophysiology of atypical depression.[20] (Fig. 1)

2.3. Neuroinflammatory system

It has been suggested that dysregulation of the immune system, including the cytokine network, is associated with the etiology and pathophysiology of depression.[47, 48] Peripheral cytokines can communicate with brain cells by various mechanisms. Many studies have suggested that imbalances in the cytokine network are associated with the pathophysiology of depression.[49, 50] (please be consistent with your citation system)

There have been many studies suggesting that proinflammatory cytokines, which initiate inflammatory immune responses, are associated with depression. First, patients and animals administered IL-2 and IFN-α experience "sickness behaviors" that resembled depression: insomnia, decreased appetite, loss of interest, and fatigue.[citation?] These "sickness behaviors" improved when they were treated with antidepressants or when the cytokines were withdrawn.[51, 52] Second, patients with depression that are otherwise healthy seem to have activated inflammatory pathways, with increased pro-inflammatory cytokines, acute-phase proteins, and increased expression of chemokines and adhesion molecules. Third, chronic

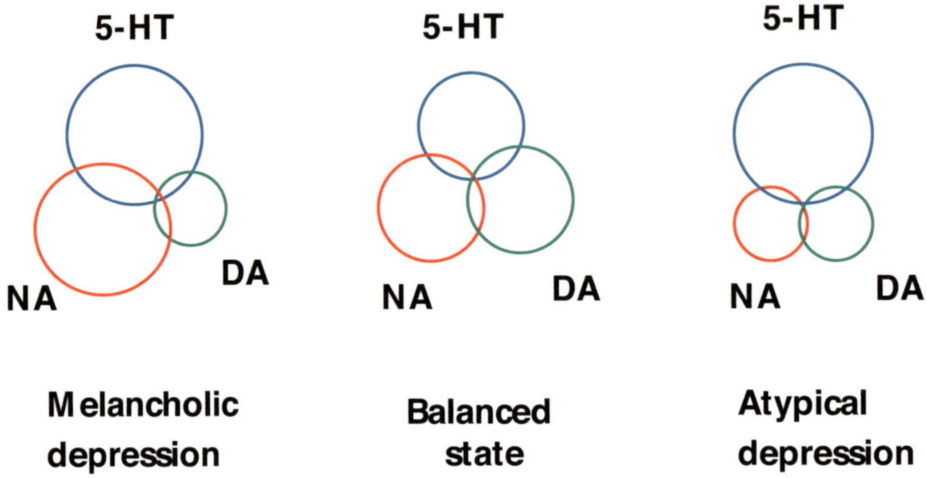

5-HT, serotonin. NA, noradrenaline. DA, dopamine.

Figure 1. Monoamine hypothesis of depression subtype. A simplistic and hypothetical model show that monoamines are differently affected in atypical and melancholic depressions and that monoaminergic neurotransmission is 'out of tune', rather than deficient. The circles represent the increased or decreased monoaminergic functioning and capacity.

inflammatory diseases such as multiple sclerosis and rheumatoid arthritis, are frequently accompanied by depression (See the references in the introduction of [47]).

The pro-inflammatory cytokines have been found to have profound effects on the metabolism of brain serotonin, dopamine, and noradrenaline in mice and rats [53]. Indeed, the activation of inflammatory pathways within the brain is believed to contribute to a confluence of decreased neurotrophic support and altered glutamate release/reuptake, as well as oxidative stress, leading to excitotoxicity and loss of glial elements, consistent with neuropathologic findings that characterize depressive disorders.[48] A recent meta-analysis convincingly suggested that IL-6 and TNF-alpha levels are elevated in depressive patients.[54]

Adipocytes, the source of leptin, also produce cytokines, such as TNF-α and IL-6. Indeed, in obese subjects, it has been estimated that about 30% of the circulating IL-6 is derived from adipose tissue.[55] However, the effects of leptin are generally opposite those of the pro-inflammatory cytokines, and include the induction of anorexia, anhedonia, and increased sympathetic nervous system activity.[56]

The immune system plays an important role in the regulation of leptin production.[79] This communication between the immune and adipose systems is bidirectional, since leptin in turn is involved in the regulation of immune responses. Indeed, leptin regulates pro-inflammatory immune responses, by up-regulating both phagocytosis and the production of pro-inflammatory cytokines. Moreover, leptin deficiency is accompanied by an increased susceptibility to endotoxin-induced lethality and a decreased induction of anti-inflammatory cytokines in rodents,[57] thus further suggesting close connections between leptin and the immune system.

Hypersomnia is one of the main symptoms of atypical depression. Cytokines are important sleep regulatory substances among many factors that are involved in sleep regulation. Among cytokines, interleukin IL-1 and TNF-α have been determined to be important sleep-promoting substances. Early studies in humans have shown that sleep onset is associated with the increased activity of IL-1, followed by elevations of IL-2 that appeared to be related to a decline in plasma cortisol and the appearance of slow wave sleep.[58] IL-4, one of the anti-somnogenic cytokines, inhibits the production or release of other substances implicated in sleep regulation, such as nuclear factor kappa B.

Atypical depression has been linked to decreased IL-4 and increased IL-2 compared to individuals without atypical features in one study,[59] while another study reported decreased IL-2 in atypical depression compared with controls.[60] Individuals with atypical depression had significantly higher levels of inflammatory markers than persons with melancholic depression and controls.[61] Overall, findings on inflammatory markers among those with melancholic versus atypical depression have been contradictory. Taking into consideration a meta-analysis that body mass index may interact with C-reactive protein and IL-6 to yield a potential tri-directional relationship between adiposity, inflammation and depression,[62] the high BMI levels of those with atypical depression may indicate a differential association between atypical depression with inflammation compared with melancholic depression, as was also postulated.[32]

2.4. Neuroplasicity

A time-lag in clinical response after the administration of an antidepressant drug suggests that alterations in monoamine metabolism alone cannot explain the entire antidepressant effect. In this respect, it was suggested that the mechanism of action of antidepressant drugs may be associated with intracellular signal transduction pathways that are linked to the expression of specific genes.[63]

The neuroplasticity hypothesis proposes that depression results from an inability to make the appropriate neuronal proliferation in response to stress.[64] Brain-derived neurotrophic factor (BDNF), an important member of the neurotrophin family, is a key component of the neuroplasticity hypothesis. The molecule acts on neurons at both presynaptic and postsynaptic sites by binding to its tyrosine kinase receptor (TrkB), resulting in the internalization of the BDNF TrkB complex-signaling endosome. [65]

A growing body of evidence shows similar results through the direct measurements of BDNF in the serum and plasma. [66-68] Antidepressants lead to the up-regulation of the cAMP response element-binding (CREB) protein and an increase in the expression of neurotrophic factors through their stimulation of intracellular pathways. By taking antidepressants, depressed patients increase their serum BDNF levels close to the physiological level. [69-71] Furthermore, studies show that the enhancement of BDNF expression may be an important element in the clinical response to antidepressant treatment. [72] The BDNF molecule has been shown to likely contribute to the "final common pathway" for different antidepressant approaches. The various antidepressant approaches include: antidepressants [73], electroconvulsive therapy [73, 74], exercise [75, 76], and repetitive transcranial magnetic stimulation. [77]

A meta-analysis including 1504 subjects indicated that BDNF levels increased significantly after antidepressant treatment and that there was a significant correlation between changes in BDNF levels and depression scores. The researchers also found a difference between pre-treatment patients and healthy controls and a small, but significant, difference between treated patients and healthy controls. [78]

Low serum BDNF levels have been found in depressed patients, however, no study has systematically investigated whether depression subtypes contribute to the low BDNF levels found in depressed subjects. One study including 1070 patients with a diagnosis of major depressive disorder within the past 6-month diagnosis from the Netherlands Study of Depression and Anxiety (NESDA) was reported. The Composite International Diagnostic Interview (CIDI) and Inventory of Depressive Symptoms (IDS) items were tested individually in separate multiple regression analyses with serum BDNF level as the dependent variable and the CIDI or IDS item as an independent variable. Subsequently, BDNF levels were compared between patients with seasonal affective disorder (based on the Seasonal Pattern Assessment Questionnaire) and melancholic depression, atypical depression, and moderate depression (based on a latent class analysis). Serum BDNF levels did not significantly differ between patients with melancholic depression, atypical depression, and moderate depression.[79]

Another study with same subjects (NESDA) examined the association between serum levels of BDNF and plasma levels of IL-6 and TNF-α in patients with MDD (n = 1070) and non-depressed controls (n = 379). Multiple regression analyses with serum BDNF as the dependent variable was used and the presence of BDNF–cytokine associations in DSM-IV-assigned melancholic MDD patients was tested. Stratified analyses showed that BDNF levels are indeed positively associated with IL-6 levels in MDD patients, but not in non-depressed controls. When further stratified for melancholic and non-melancholic MDD, IL-6 emerged as a robust positive predictor of BDNF only in the melancholic sample, wherein serum BDNF levels were accordingly enhanced. Post-hoc exploratory analyses verified an accentuated positive association of BDNF levels with leucocyte counts in melancholia. No significant associations emerged between BDNF and TNF-α.[80] Another study found that IL-6 and TNF-α specifically enhanced BDNF secretion in monocytes, whereas typical Th1- and Th2-cytokines did not show any effect on monocytes. Otherwise, only IL-6 and tumor necrosis factor-alpha (TNF-α) were found to have the ability to enhance extracellular BDNF levels in human monocytes. Intriguingly, levels of BDNF in antidepressant-free melancholics – the group presenting with the most clear-cut BDNF–IL-6 association – was not significantly different from both non-melancholics and controls, suggesting that low serum BDNF may not be a hallmark of melancholia.[81]. This finding is concordant with a recent study showing that serum BDNF levels of antidepressant-free melancholic patients are not different from healthy controls [82].

Although BDNF is believed to be transported over the blood – brain barrier [83], and significant correlations have been found between peripheral BDNF and measures of central neuroplasticity [84], we cannot be sure that measuring serum BDNF reflects the brain expression of BDNF adequately. Currently, however, measuring BDNF in the peripheral blood is the only feasible method as other methods would be far more invasive.

Conclusively, these few studies suggest that there is not much possibility of different neuroplastic mechanisms between atypical and melancholic depression.

2.5. Neuroimaging factor

Recent neuroimaging studies have focused on the neurobiological differences between healthy controls and abnormalities associated with MDD, such as dysfunctional or structural differences in cerebral regions, including the prefrontal cortex, amygdala, anterior cingulate cortex (ACC), and hippocampus. [85-88]

Regional CBF and metabolism are consistently increased in the amygdala, orbital cortex, and medial thalamus, and decreased in the dorsomedial/dorsal anterolateral PFC and anterior cingulate cortex ventral to the genu of the corpus callosum (subgenual PFC) on positron emission tomography (PET) imaging studies in MDD subjects without medication, as compared to healthy controls. [89, 90] These circuits have also been implicated more generally in emotional behavior.

Previous structural magnetic resonance imaging (MRI) studies using region-of-interest (ROI) analyses have shown a variety of inconsistent findings. [91, 92] These inconsistencies can likely be explained by variability in the ROI criteria between studies and inconsistency in ROI validation.[91, 93, 94] Consequently, voxel-based morphometry (VBM)[95] is being increasingly used as a viable alternative methodology for detecting structural abnormalities in patients with neuropsychiatric disorders, including MDD.[96-99] Previous MDD VBM studies have also shown reduced gray matter density in the hippocampus. [97, 98, 100] Recently, it has been reported that the gray matter density of several regions associated with emotion regulation, particularly dorsal raphe nucleus, was lower in MDD patients.[101]

Because depression is heterogeneous, subtyping the disease will be helpful for understanding imaging results. However, there are few imaging studies which were done according to depression subtype. There is no VBM study.

One chimeric faces study measured of perceptual asymmetry and showed that those with atypical depression differed from those with typical depression and controls in showing abnormally large right hemisphere bias. A chimeric face consists of fusion of a neutral right half-face with a smiling left half-face. Its mirror image (creating a neutral left half-face fused with a smiling right half-face) is randomly placed above or below. The task is to quickly determine which of the two faces is happier. Preference for choosing one side as happier relative to the other has been interpreted as reflecting increased activation of the contralateral parietal lobe,[102] although inhibitory mechanisms could also be hypothesized. This was present in patients having either MDD or dysthymia and was not related to anxiety, physical anhedonia, or vegetative symptoms. In contrast, patients with melancholic depression showed essentially no right hemisphere bias. The authors suggest that this is further evidence that atypical depression is a biologically-distinct subtype and underscores the importance of this diagnostic distinction for neurophysiologic studies.[103]

Single photon emission computerized tomography (SPECT) in 50 depressed patients with MDD, including subtype assessment indicated differential brain activity in patients with atypical depression compared with typical depression. [104] Patients with melancholic depression (N=16) and patients with undifferentiated depression (N=20) each differed from controls (N=20) in 10 brain regions, but did not differ from each other in any of the 17 regions. In contrast, patients with atypical depression (N=14) differed from patients with melancholic depression in nine regions and from patients with undifferentiated depression in 10 regions, while showing differences from controls in five brain regions. In two brain regions, patients with atypical depression differed from both controls and at least one of the other depressed groups (Table 1). Conclusively, those with atypical depression had increased frontal, temporal, and parietal perfusion coupled with decreased occipital perfusion, relative to the other two depressed groups. Patients with atypical depression also had increased right frontal perfusion, whereas those with melancholia and undifferentiated depression had decreased perfusion in the majority of nonoccipital regions, relative to controls. Thus, all three depressed groups showed abnormal perfusion, but the patterns differed. Melancholia and undifferentiated depression had similar patterns of abnormal perfusion that differed from those with atypical depression.

Region	M vs. C	U vs. C	A vs. C	A vs. M	A vs. U	M vs. U
Brain stem						
Right frontal			↑	↑	↑	
Left frontal	↓			↑		
Right parietal	↓	↓	↓			
Left parietal	↓	↓	↓			
Right medial temporal	↓	↓		↑	↑	
Left medial temporal		↓		↑	↑	
Right lateral temporal	↓	↓		↑	↑	
Left lateral temporal	↓	↓		↑	↑	
Right occipital				↓	↓	
Left occipital		↓	↓	↓		
Right thalamus					↑	
Left thalamus	↓		↓			
Right globus pallidus						
Left globus pallidus	↓	↓			↑	
Right caudate	↓	↓			↑	
Left caudate	↓	↓		↑	↑	

A, atypical depression. C, control. M, melancholic depression. U, undifferentiated depression.

↑ , first group is significantly increased relative to the second group.

↓ , first group is significantly decreased relative to the second group.

Table 1. Perfusion findings of depressed patients and controls [129], adapted from [105].

These imaging studies are consistent, suggesting that atypical depression does not have the biological features of melancholia.

3. Conclusions

Major depressive disorder is considered to be a clinically heterogeneous disorder and the diagnosis is based on a patient's symptoms, not on any laboratory tests. Consequently, MDD's pathophysiology is unsettled. Currently, researchers have determined that MDD results from the interaction of multiple genetic factors and various environmental factors, such as childhood adversity and many stressful life events. Although the development of antidepressant drugs has skyrocketed in recent decades, the neurobiological effects underlying the therapeutic actions of these agents remain poorly understood. Considering the biological mechanism of depressive subtypes, it is helpful to understand the pathogenesis of each depressive disorder in order to predict an individual's response to treatment for MDD. For example, melancholic depression is associated with hyperactivity of the HPA axis while atypical depression is associated with hypoactivity of the HPA axis. Researchers have searched for biological mechanisms according to depression subtypes in an effort tto understand the pathogenesis of depression subtypes.

Concerning pharmacological treatment, it was reported that the group of patients with atypical depression showed a significantly higher cortisol response to desipramine, a relatively selective noradrenaline reuptake inhibitor, than a group with no atypical symptoms and a group with mood reactivity as the only atypical symptom, indicating that atypical depression may be associated with a smaller impairment of the noradrenaline neurotransmitter system. Similarly, hypersecretion of corticotropin-releasing hormone (CRH) and the resulting hypercortisolism were not found in patients with atypical depression. Imaging studies are consistent with that finding, suggesting that atypical depression does not have the biological features of melancholia.

The results are summarized in Table 2.

Melancholic depression	Atypical depression
DSM-5 subsymptoms per subtype[5]:	
A. One of the following:	A. Mood reactivity
1. Loss of pleasure	
2. Lack of reactivity to usually pleasurable stimuli	
B. Three of following:	B. Two of the following:
1. Depressed mood with profound despondency, despair , moroseness	1. Weight gain/increase in appetite
	2. Hypersomnia
2. Symptoms at worst in morning	3. Leaden paralysis
3. Early-morning awakening	4. Interpersonal rejection sensitivity

Melancholic depression	Atypical depression
4. Psychomotor agitation/retardation 5. Anorexia or weight loss 6. Guilt	C. Criteria are not met for "with melancholic features" or "with catatonia" during the same episode.
Neurobiological mechanisms per subtype:	
Increased sympathetic activity	Decreased sympathetic activity
Hyperactive HPA axis	Hypoactive HPA axis
Activated CRF system	CRF deficiency
Increased susceptibility for infection	Increased susceptibility for inflammation
Low BDNF	Low BDNF
	Increased right frontal or parietal region in imaging studies

DSM, Diagnostic and Statistical Manual of Mental Disorders.

Table 2. Different clinical symptoms and biological mechanisms between melancholic and atypical depression based on the DSM-5

Acknowledgements

This work was supported by a grant of Soonchunhyang University.

Author details

Hwa-Young Lee[1*] and Yong-Ku Kim[2]

*Address all correspondence to: leehway@gmail.com

1 Department of Psychiatry, College of Medicine, Soonchunhyang University, Republic of Korea

2 Department of Psychiatry, College of Medicine, Korea University, Republic of Korea

References

[1] Bierut, L.J., et al., *Major depressive disorder in a community-based twin sample: are there different genetic and environmental contributions for men and women?* Arch Gen Psychiatry, 1999. 56(6): p. 557-63.

[2] Hamet, P. and J. Tremblay, *Genetics and genomics of depression.* Metabolism, 2005. 54(5 Suppl 1): p. 10-5.

[3] Murray, C.J. and A.D. Lopez, *Global mortality, disability, and the contribution of risk factors: Global Burden of Disease Study.* Lancet, 1997. 349(9063): p. 1436-42.

[4] Caspi, A., et al., *Influence of life stress on depression: moderation by a polymorphism in the 5-HTT gene.* Science, 2003. 301(5631): p. 386-9.

[5] American Psychiatric Association, *Diagnostic and statistical manual of mental disorders: DSM-5, 5th edition. Washington DC: American Psychiatric Association.* 2013.

[6] Thase, M.E., *Recognition and diagnosis of atypical depression.* J Clin Psychiatry, 2007. 68 Suppl 8: p. 11-6.

[7] Holsboer, F., *The rationale for corticotropin-releasing hormone receptor (CRH-R) antagonists to treat depression and anxiety.* J Psychiatr Res, 1999. 33(3): p. 181-214.

[8] Holsboer, F., *The corticosteroid receptor hypothesis of depression.* Neuropsychopharmacology, 2000. 23(5): p. 477-501.

[9] Holsboer, F. and N. Barden, *Antidepressants and hypothalamic-pituitary-adrenocortical regulation.* Endocr Rev, 1996. 17(2): p. 187-205.

[10] Nemeroff, C.B., *The corticotropin-releasing factor (CRF) hypothesis of depression: new findings and new directions.* Mol Psychiatry, 1996. 1(4): p. 336-42.

[11] De Kloet, E.R., et al., *Brain corticosteroid receptor balance in health and disease.* Endocr Rev, 1998. 19(3): p. 269-301.

[12] Meijer, O.C. and E.R. de Kloet, *Corticosterone and serotonergic neurotransmission in the hippocampus: functional implications of central corticosteroid receptor diversity.* Crit Rev Neurobiol, 1998. 12(1-2): p. 1-20.

[13] Wolkowitz, O.M., et al., *Ketoconazole administration in hypercortisolemic depression.* Am J Psychiatry, 1993. 150(5): p. 810-2.

[14] Holsboer, F., et al., *Altered hypothalamic-pituitary-adrenocortical regulation in healthy subjects at high familial risk for affective disorders.* Neuroendocrinology, 1995. 62(4): p. 340-7.

[15] Modell, S., et al., *Hormonal response pattern in the combined DEX-CRH test is stable over time in subjects at high familial risk for affective disorders.* Neuropsychopharmacology, 1998. 18(4): p. 253-62.

[16] Pariante, C.M. and A.H. Miller, *Glucocorticoid receptors in major depression: relevance to pathophysiology and treatment.* Biol Psychiatry, 2001. 49(5): p. 391-404.

[17] Arborelius, L., et al., *The role of corticotropin-releasing factor in depression and anxiety disorders.* J Endocrinol, 1999. 160(1): p. 1-12.

[18] Nemeroff, C.B. and M.J. Owens, *Treatment of mood disorders.* Nat Neurosci, 2002. 5 Suppl: p. 1068-70.

[19] Nemeroff, C.B., *The role of corticotropin-releasing factor in the pathogenesis of major depression.* Pharmacopsychiatry, 1988. 21(2): p. 76-82.

[20] Lee, H., et al., *Association of glucocorticoid receptor polymorphisms with the susceptibility to major depressive disorder and treatment responses in Korean depressive patients.* Acta Neuropsychiatrica, 2009. 21(1): p. 11-17.

[21] Binder, E.B., et al., *Association of polymorphisms in genes regulating the corticotropin-releasing factor system with antidepressant treatment response.* Arch Gen Psychiatry, 2010. 67(4): p. 369-79.

[22] Binder, E.B., et al., *Polymorphisms in FKBP5 are associated with increased recurrence of depressive episodes and rapid response to antidepressant treatment.* Nat Genet, 2004. 36(12): p. 1319-25.

[23] Claes, S., *Glucocorticoid receptor polymorphisms in major depression.* Ann N Y Acad Sci, 2009. 1179: p. 216-28.

[24] Lightman, S.L., *The neuroendocrinology of stress: a never ending story.* J Neuroendocrinol, 2008. 20(6): p. 880-4.

[25] Dinan, T.G., et al., *Further neuroendocrine evidence of enhanced vasopressin V3 receptor responses in melancholic depression.* Psychol Med, 2004. 34(1): p. 169-72.

[26] Dinan, T.G. and L.V. Scott, *Anatomy of melancholia: focus on hypothalamic-pituitary-adrenal axis overactivity and the role of vasopressin.* J Anat, 2005. 207(3): p. 259-64.

[27] Antonijevic, I.A., et al., *Sexually dimorphic effects of GHRH on sleep-endocrine activity in patients with depression and normal controls - part II: hormone secretion.* Sleep Res Online, 2000. 3(1): p. 15-21.

[28] Casper, R.C., et al., *Weight loss, cortisol levels, and dexamethasone suppression in major depressive disorder.* Acta Psychiatr Scand, 1987. 75(3): p. 243-50.

[29] Miller, K.B. and J.C. Nelson, *Does the dexamethasone suppression test relate to subtypes, factors, symptoms, or severity?* Arch Gen Psychiatry, 1987. 44(9): p. 769-74.

[30] Casper, R.C., et al., *Cortisol measures in primary major depressive disorder with hypersomnia or appetite increase.* J Affect Disord, 1988. 15(2): p. 131-40.

[31] Thase, M.E., et al., *Sleep EEG and DST findings in anergic bipolar depression.* Am J Psychiatry, 1989. 146(3): p. 329-33.

[32] Gold, P.W. and G.P. Chrousos, *Organization of the stress system and its dysregulation in melancholic and atypical depression: high vs low CRH/NE states.* Mol Psychiatry, 2002. 7(3): p. 254-75.

[33] Stetler, C. and G.E. Miller, *Depression and hypothalamic-pituitary-adrenal activation: a quantitative summary of four decades of research.* Psychosom Med, 2011. 73(2): p. 114-26.

[34] Antonijevic, I.A., *Depressive disorders -- is it time to endorse different pathophysiologies?* Psychoneuroendocrinology, 2006. 31(1): p. 1-15.

[35] McGinn, L.K., G.M. Asnis, and E. Rubinson, *Biological and clinical validation of atypical depression.* Psychiatry Res, 1996. 60(2-3): p. 191-8.

[36] Geracioti, T.D., Jr., P.T. Loosen, and D.N. Orth, *Low cerebrospinal fluid corticotropin-releasing hormone concentrations in eucortisolemic depression.* Biol Psychiatry, 1997. 42(3): p. 165-74.

[37] Stockmeier, C.A., et al., *Increase in serotonin-1A autoreceptors in the midbrain of suicide victims with major depression-postmortem evidence for decreased serotonin activity.* J Neurosci, 1998. 18(18): p. 7394-401.

[38] Lemonde, S., et al., *Impaired repression at a 5-hydroxytryptamine 1A receptor gene polymorphism associated with major depression and suicide.* J Neurosci, 2003. 23(25): p. 8788-99.

[39] Ellis, P.M. and C. Salmond, *Is platelet imipramine binding reduced in depression? A meta-analysis.* Biol Psychiatry, 1994. 36(5): p. 292-9.

[40] Bonkale, W.L., G. Turecki, and M.C. Austin, *Increased tryptophan hydroxylase immunoreactivity in the dorsal raphe nucleus of alcohol-dependent, depressed suicide subjects is restricted to the dorsal subnucleus.* Synapse, 2006. 60(1): p. 81-5.

[41] Herman, G.M., *Pharmacology of Antidepressant: Selectivity or Multiplicity?* J Clin Psychiatry, 1999. 60 suppl 17: p. 4-8.

[42] Kasper, S., *Efficacy of antidepressants in the treatment of severe depression: the place of mirtazapine.* J Clin Psychopharmacol, 1997. 17 Suppl 1: p. 19S-28S.

[43] Preskorn, S.H., *Selection of an antidepressant: mirtazapine.* J Clin Psychiatry, 1997. 58 Suppl 6: p. 3-8.

[44] Geracitano, R., et al., *On the effects of psychostimulants, antidepressants, and the antiparkinsonian drug levodopa on dopamine neurons.* Ann N Y Acad Sci, 2006. 1074: p. 320-9.

[45] Tsigos, C. and G.P. Chrousos, *Hypothalamic-pituitary-adrenal axis, neuroendocrine factors and stress.* J Psychosom Res, 2002. 53(4): p. 865-71.

[46] Owens, M.J. and C.B. Nemeroff, *Role of serotonin in the pathophysiology of depression: focus on the serotonin transporter.* Clin Chem, 1994. 40(2): p. 288-95.

[47] Kim, Y.K., et al., *Cytokine imbalance in the pathophysiology of major depressive disorder.* Prog Neuropsychopharmacol Biol Psychiatry, 2007. 31(5): p. 1044-53.

[48] Miller, A.H., V. Maletic, and C.L. Raison, *Inflammation and its discontents: the role of cytokines in the pathophysiology of major depression.* Biol Psychiatry, 2009. 65(9): p. 732-41.

[49] Kronfol, Z., *Immune dysregulation in major depression: a critical review of existing evidence.* Int J Neuropsychopharmacol, 2002. 5(4): p. 333-43.

[50] Schiepers, O.J., M.C. Wichers, and M. Maes, *Cytokines and major depression.* Prog Neuropsychopharmacol Biol Psychiatry, 2005. 29(2): p. 201-17.

[51] Bonaccorso, S., et al., *Increased depressive ratings in patients with hepatitis C receiving interferon-alpha-based immunotherapy are related to interferon-alpha-induced changes in the serotonergic system.* J Clin Psychopharmacol, 2002. 22(1): p. 86-90.

[52] Capuron, L., A. Ravaud, and R. Dantzer, *Early depressive symptoms in cancer patients receiving interleukin 2 and/or interferon alfa-2b therapy.* J Clin Oncol, 2000. 18(10): p. 2143-51.

[53] Dunn, A.J., J. Wang, and T. Ando, *Effects of cytokines on cerebral neurotransmission. Comparison with the effects of stress.* Adv Exp Med Biol, 1999. 461: p. 117-27.

[54] Dowlati, Y., et al., *A meta-analysis of cytokines in major depression.* Biol Psychiatry, 2010. 67(5): p. 446-57.

[55] Yudkin, J.S., et al., *Inflammation, obesity, stress and coronary heart disease: is interleukin-6 the link?* Atherosclerosis, 2000. 148(2): p. 209-14.

[56] Haynes, W.G., et al., *Receptor-mediated regional sympathetic nerve activation by leptin.* J Clin Invest, 1997. 100(2): p. 270-8.

[57] Faggioni, R., et al., *Leptin deficiency enhances sensitivity to endotoxin-induced lethality.* Am J Physiol, 1999. 276(1 Pt 2): p. R136-42.

[58] Moldofsky, H., et al., *The relationship of interleukin-1 and immune functions to sleep in humans.* Psychosom Med, 1986. 48(5): p. 309-18.

[59] Yoon, H.K., et al., *Role of cytokines in atypical depression.* Nord J Psychiatry, 2012. 66(3): p. 183-8.

[60] Anisman, H., et al., *Endocrine and cytokine correlates of major depression and dysthymia with typical or atypical features.* Mol Psychiatry, 1999. 4(2): p. 182-8.

[61] Lamers, F., et al., *Evidence for a differential role of HPA-axis function, inflammation and metabolic syndrome in melancholic versus atypical depression.* Mol Psychiatry, 2013. 18(6): p. 692-9.

[62] Howren, M.B., D.M. Lamkin, and J. Suls, *Associations of depression with C-reactive protein, IL-1, and IL-6: a meta-analysis.* Psychosom Med, 2009. 71(2): p. 171-86.

[63] Duman, R.S., et al., *Neuronal plasticity and survival in mood disorders.* Biol Psychiatry, 2000. 48(8): p. 732-9.

[64] Popoli, M., M. Gennarelli, and G. Racagni, *Modulation of synaptic plasticity by stress and antidepressants.* Bipolar Disord, 2002. 4(3): p. 166-82.

[65] Lu, B., *BDNF and activity-dependent synaptic modulation*. Learn Mem, 2003. 10(2): p. 86-98.

[66] Karege, F., et al., *Decreased serum brain-derived neurotrophic factor levels in major depressed patients*. Psychiatry Res, 2002. 109(2): p. 143-8.

[67] Karege, F., et al., *Low brain-derived neurotrophic factor (BDNF) levels in serum of depressed patients probably results from lowered platelet BDNF release unrelated to platelet reactivity*. Biol Psychiatry, 2005. 57(9): p. 1068-72.

[68] Shimizu, E., et al., *Alterations of serum levels of brain-derived neurotrophic factor (BDNF) in depressed patients with or without antidepressants*. Biol Psychiatry, 2003. 54(1): p. 70-5.

[69] Aydemir, O., A. Deveci, and F. Taneli, *The effect of chronic antidepressant treatment on serum brain-derived neurotrophic factor levels in depressed patients: a preliminary study*. Prog Neuropsychopharmacol Biol Psychiatry, 2005. 29(2): p. 261-5.

[70] Gervasoni, N., et al., *Partial normalization of serum brain-derived neurotrophic factor in remitted patients after a major depressive episode*. Neuropsychobiology, 2005. 51(4): p. 234-8.

[71] Gonul, A.S., et al., *Effect of treatment on serum brain-derived neurotrophic factor levels in depressed patients*. Eur Arch Psychiatry Clin Neurosci, 2005. 255(6): p. 381-6.

[72] Russo-Neustadt, A., et al., *Physical activity-antidepressant treatment combination: impact on brain-derived neurotrophic factor and behavior in an animal model*. Behav Brain Res, 2001. 120(1): p. 87-95.

[73] Nibuya, M., S. Morinobu, and R.S. Duman, *Regulation of BDNF and trkB mRNA in rat brain by chronic electroconvulsive seizure and antidepressant drug treatments*. J Neurosci, 1995. 15(11): p. 7539-47.

[74] Altar, C.A., et al., *Effects of electroconvulsive seizures and antidepressant drugs on brain-derived neurotrophic factor protein in rat brain*. Biol Psychiatry, 2003. 54(7): p. 703-9.

[75] Oliff, H.S., et al., *Exercise-induced regulation of brain-derived neurotrophic factor (BDNF) transcripts in the rat hippocampus*. Brain Res Mol Brain Res, 1998. 61(1-2): p. 147-53.

[76] Garcia, C., et al., *The influence of specific noradrenergic and serotonergic lesions on the expression of hippocampal brain-derived neurotrophic factor transcripts following voluntary physical activity*. Neuroscience, 2003. 119(3): p. 721-32.

[77] Muller, M.B., et al., *Long-term repetitive transcranial magnetic stimulation increases the expression of brain-derived neurotrophic factor and cholecystokinin mRNA, but not neuropeptide tyrosine mRNA in specific areas of rat brain*. Neuropsychopharmacology, 2000. 23(2): p. 205-15.

[78] Brunoni, A.R., M. Lopes, and F. Fregni, *A systematic review and meta-analysis of clinical studies on major depression and BDNF levels: implications for the role of neuroplasticity in depression*. Int J Neuropsychopharmacol, 2008. 11(8): p. 1169-80.

[79] Bus, B.A., et al., *Low serum BDNF levels in depressed patients cannot be attributed to individual depressive symptoms or symptom cluster.* World J Biol Psychiatry, 2014. 15(7): p. 561-9.

[80] Patas, K., et al., *Association between serum brain-derived neurotrophic factor and plasma interleukin-6 in major depressive disorder with melancholic features.* Brain Behav Immun, 2014. 36: p. 71-9.

[81] Schulte-Herbruggen, O., et al., *Tumor necrosis factor-alpha and interleukin-6 regulate secretion of brain-derived neurotrophic factor in human monocytes.* J Neuroimmunol, 2005. 160(1-2): p. 204-9.

[82] Kotan, Z., et al., *Serum brain-derived neurotrophic factor, vascular endothelial growth factor and leptin levels in patients with a diagnosis of severe major depressive disorder with melancholic features.* Ther Adv Psychopharmacol, 2012. 2(2): p. 65-74.

[83] Karege, F., M. Schwald, and M. Cisse, *Postnatal developmental profile of brain-derived neurotrophic factor in rat brain and platelets.* Neurosci Lett, 2002. 328(3): p. 261-4.

[84] Lang, U.E., et al., *Correlation between serum brain-derived neurotrophic factor level and an in vivo marker of cortical integrity.* Biol Psychiatry, 2007. 62(5): p. 530-5.

[85] Canli, T., et al., *Brain activation to emotional words in depressed vs healthy subjects.* Neuroreport, 2004. 15(17): p. 2585-8.

[86] Frodl, T., et al., *Hippocampal and amygdala changes in patients with major depressive disorder and healthy controls during a 1-year follow-up.* J Clin Psychiatry, 2004. 65(4): p. 492-9.

[87] Henriques, J.B., Davidson, R.J., *Decreased responsiveness to reward in depression.* COGNITION AND EMOTION, 2000. 15(5): p. 711-724.

[88] MacQueen, G.M., et al., *Course of illness, hippocampal function, and hippocampal volume in major depression.* Proc Natl Acad Sci U S A, 2003. 100(3): p. 1387-92.

[89] Drevets, W.C., *Neuroimaging and neuropathological studies of depression: implications for the cognitive-emotional features of mood disorders.* Curr Opin Neurobiol, 2001. 11(2): p. 240-9.

[90] Manji, H.K., W.C. Drevets, and D.S. Charney, *The cellular neurobiology of depression.* Nat Med, 2001. 7(5): p. 541-7.

[91] Konarski, J.Z., et al., *Volumetric neuroimaging investigations in mood disorders: bipolar disorder versus major depressive disorder.* Bipolar Disord, 2008. 10(1): p. 1-37.

[92] Lorenzetti, V., et al., *Structural brain abnormalities in major depressive disorder: a selective review of recent MRI studies.* J Affect Disord, 2009. 117(1-2): p. 1-17.

[93] Goldstein, J.M., et al., *Cortical abnormalities in schizophrenia identified by structural magnetic resonance imaging.* Arch Gen Psychiatry, 1999. 56(6): p. 537-47.

[94] Wolkin, A., et al., *Structural magnetic resonance image averaging in schizophrenia.* Am J Psychiatry, 1998. 155(8): p. 1064-73.

[95] Ashburner, J. and K.J. Friston, *Voxel-based morphometry--the methods.* Neuroimage, 2000. 11(6 Pt 1): p. 805-21.

[96] Leung, K.K., et al., *Neural correlates of attention biases of people with major depressive disorder: a voxel-based morphometric study.* Psychol Med, 2009. 39(7): p. 1097-106.

[97] Shah, P.J., et al., *Cortical grey matter reductions associated with treatment-resistant chronic unipolar depression. Controlled magnetic resonance imaging study.* Br J Psychiatry, 1998. 172: p. 527-32.

[98] Vasic, N., et al., *Gray matter reduction associated with psychopathology and cognitive dysfunction in unipolar depression: a voxel-based morphometry study.* J Affect Disord, 2008. 109(1-2): p. 107-16.

[99] Cheng, Y.Q., et al., *Brain volume alteration and the correlations with the clinical characteristics in drug-naive first-episode MDD patients: a voxel-based morphometry study.* Neurosci Lett, 2010. 480(1): p. 30-4.

[100] Bergouignan, L., et al., *Can voxel based morphometry, manual segmentation and automated segmentation equally detect hippocampal volume differences in acute depression?* Neuroimage, 2009. 45(1): p. 29-37.

[101] Lee, H.Y., et al., *Demonstration of decreased gray matter concentration in the midbrain encompassing the dorsal raphe nucleus and the limbic subcortical regions in major depressive disorder: an optimized voxel-based morphometry study.* J Affect Disord, 2011. 133(1-2): p. 128-36.

[102] Heller, W., *Gender differences in depression: perspectives from neuropsychology.* J Affect Disord, 1993. 29(2-3): p. 129-43.

[103] Bruder, G.E., et al., *Atypical depression: enhanced right hemispheric dominance for perceiving emotional chimeric faces.* J Abnorm Psychol, 2002. 111(3): p. 446-54.

[104] Fountoulakis, K.N., et al., *The relationship of regional cerebral blood flow with subtypes of major depression.* Prog Neuropsychopharmacol Biol Psychiatry, 2004. 28(3): p. 537-46.

[105] Stewart, J.W., et al., DSM-IV depression with atypical features: is it valid? Neuropsychopharmacology, 2009. 34(13): p. 2625-32.

Chapter 2

Neurotrophic Factors and Major Depressive Disorder

Xiaobin Zhang, Jin Li, Weiwei Sha and Ru Bu

Additional information is available at the end of the chapter

http://dx.doi.org/10.5772/59373

1. Introduction

Major depressive disorder (MDD) is a serious disorder that affects approximately 17% of the population at some point in life, resulting in major social and economic consequences [1]. There is still very little known about the neurobiological alterations that underlie the pathophysiology or treatment of MDD. There is increasing evidence suggesting that brain-derived neurotrophic factor (BDNF), neurotrophin-3, and fibroblast growth factor (FGF) systems are altered in different tissue samples, including post-mortem brain tissue, cerebrospinal fluid, and blood from patients with MDD. Neurotrophins are a family of secreted growth factors that regulate survival, growth, differentiation and maintenance of neurons in both the central nervous system and the peripheral nervous system [2] and their reduced availability can result in increased cellular vulnerability or even cell death. It has been postulated that the enhanced and prolonged secretion of neurotrophic factors in response to antidepressant treatment could promote neuronal survival and protect neurons from the damaging effects of stress. These studies have led to the formulation of the neurotrophic hypothesis of depression, which proposes that reduced neurotrophic factors levels predispose to depression, whereas increases neurotrophic factors produce an antidepressant action.

2. Brain-Derived Neurotrophic Factor (BDNF) in Major Depressive Disorder (MDD)

The *BDNF* gene is located on chromosome 11p13 and is synthesized as the precursor pre-pro-BDNF that is cleaved into pro-BDNF, which is then further cleaved into the 27-kDa mature protein [3]. BDNF plays an important role in the survival, differentiation, and outgrowth of various neurons in the peripheral and central nervous systems during development [2]. It is also involved in nerve regeneration and maintenance of structural integrity and neuronal plasticity in the adult brain, which includes the regulation of synaptic activity and neurotransmitter synthesis [4]. Thus, a pathological alteration of BDNF level can lead to defects in neuronal maintenance and neural regeneration as well as structural abnormalities and reduced plasticity that can impair an individual's ability to adapt to traumatic situations. Indeed, many pre-clinical and clinical studies implicate the modulation of BDNF expression in the behavioral manifestations of depression [5].

BDNF has shown antidepressant-like effects in animal models of depression. For instance, in the forced swim test and learned helplessness models of depression, BDNF infusion into the midbrain alleviated depressive behavior [6]. Moreover, a single bilateral infusion of BDNF into the dentate gyrus (DG) of the hippocampus had an antidepressant effect in both paradigms [7].

Various stressors can suppress *BDNF* mRNA and protein levels in different brain regions particularly in the hippocampus [8], which is associated with the development of depressive behavior. However, other reports found the contrary or else no changes in BDNF expression in animal models of depression and antidepressant efficacy [9-10]. Acute and chronic stress also increase hippocampal *BDNF* mRNA and protein levels [11-12], and it has been speculated that this could serve as a protective mechanism to offset the destructive effects of stress in the hippocampus.

Several classes of antidepressant increase BDNF level in the healthy rodent brain with chronic treatment, and also reverse stress-induced downregulation of BDNF [5]. Thus, regardless of their primary mechanism of action, antidepressants share the ability to rapidly activate signaling through the TrkB receptor and induce a long-lasting increase in BDNF production [13]. However, they fail to produce any behavioral changes in transgenic mice with reduced BDNF levels or TrkB signaling, whereas wild-type mice exhibit a normal behavioral response [14], demonstrating that BDNF release and TrkB signaling are not only sufficient but also necessary for antidepressant-like behavioral effects.

Electroconvulsive seizures increase mRNA levels of both *BDNF* and *TrkB* in the rat hippocampus, and chronic seizures blocked the downregulation of *BDNF* transcript level in the hippocampus in response to restraint-induced stress [15]. Repetitive transcranial magnetic stimulation (rTMS) is increasingly being used as a therapeutic tool to treat depressive disorder, based on reports that rTMS increases *BDNF* mRNA and protein in the rat hippocampus [16], similar to effects observed after administration of antidepressants or electroconvulsive seizures, which suggest that rTMS and antidepressants likely share a common molecular mechanism of action.

Several studies of post-mortem brain samples have implicated BDNF as a factor in the pathophysiology of depressive disorders; in one report, BDNF and TrkB levels were reduced in the hippocampus of patients with depressive disorder [17]. Given that BDNF expression is downregulated in response to stress, structural changes in the hippocampus of depressed individuals may be attributed in part to reductions in BDNF and TrkB levels [18]. In patients with depression, a decrease in prefrontal cortical volume is correlated with decreased BDNF and TrkB levels [19]. These findings indicate that depression affects BDNF expression in limbic regions. As observed in animal models, postmortem tissue samples from human subjects show increased BDNF levels in the hippocampus and cortex after long-term antidepressant use as compared to untreated subjects [20].

The Val66Met polymorphism is located in the pro-BDNF sequence, which is cleaved post-transcriptionally; while not affecting mature BDNF protein function, it has been shown to alter activity-dependent BDNF secretion in cultured cells [21], as well as intracellular distribution, packaging, and release of the BDNF protein in vitro [22]. Interestingly, mutant mice carrying the Val66Met polymorphism show reduced BDNF secretion but no differences in the level of total BDNF [21]. Moreover, hippocampal volume is reduced in mice with the BDNF Met/Met or Val/Met allele as compared to wild-type mice. Our research found that the presence of the Val66Met polymorphism was not significantly correlated with serum BDNF levels in depressive patients nor in control subjects. Postmortem analyses and imaging studies have found that depressive patients have significantly smaller hippocampal volumes compared to controls, which is reversed by antidepressant treatment [23]; smaller volumes were also observed in both patients and controls carrying the Met-BDNF allele as compared to individuals that were homozygous for the Val-BDNF allele. It was therefore concluded that Met-BDNF allele carriers may be susceptible to depression due to smaller hippocampal volumes [24]. Our laboratory has also demonstrated that the Val66Met allele frequency was similar among depressed and non-depressed subjects, consistent with previous studies in Asian populations [25-26]; however, an association between the BDNF Val66Met polymorphism and depression has been reported in a geriatric Chinese population [27]. A significant genetic association between the BDNF Val66Met polymorphism and treatment response in depressed patients was detected in a meta-analysis demonstrating that Val/Met heterozygous patients had a better response rate than Val/Val homozygous patients, especially among Asians [28]. Moreover, Met allele carriers showed a favorable response to antidepressant treatment [29]. However, other studies have not found any link between genetic variation in *BDNF* and antidepressant treatment response or remission [30-31]. The results from these studies indicate that the association between BDNF polymorphisms and the manifestation of depressive disorder symptoms or antidepressant efficacy remains controversial.

Suicide is a major public health problem linked to depression. *BDNF* and *TrkB* mRNA and BDNF protein levels were significantly reduced in both the prefrontal cortex and hippocampus of suicidal as compared to non-suicidal subjects [32]. BDNF downregulation was also observed in the hippocampus and prefrontal cortex of drug-free suicidal patients as compared to non-suicidal controls. In addition, suicidal subjects receiving antidepressant treatment showed no changes BDNF level, suggesting that psychotropic drugs counter the decrease in BDNF associated with depression [33]. DNA methylation at specific CpG sites in the *BDNF* promoter/exon IV was detected when postmortem brains samples from suicidal subjects were compared to those of non-suicidal controls. These findings support a role for the BDNF pathway in suicidal behavior associated with depression.

BDNF is present in human blood, although it is more highly concentrated in brain tissue. It was previously reported that BDNF could cross the blood-brain barrier, and that serum and brain BDNF levels showed similar changes during aging in rats, suggesting that the former is a reflection of the latter [34]. Serum BDNF level was markedly lower in depressed than in healthy control subjects and was negatively correlated with depression severity, an effect that was more pronounced in females [35]. Similar findings were reported in another study, which also found greater changes in serum BDNF protein level in female but not in male patients treated with antidepressants during a 4-week period [36]. Some studies report that serum BDNF level is negatively correlated with depression severity [37-38]. Other investigators have found that plasma BDNF level is positively correlated with scores on the Hamilton Depression Rating Scale [39], although our own research did not substantiate these findings [40]. Serum BDNF level is determined by at least eight independent factors: time of blood withdrawal, time of storage, food intake before sampling, urbanicity, age, sex, smoking status, and drinking behavior [41]. However, the conflicting data from various studies suggest that other factors are likely to modulate BDNF level in depression.

3. Glial cell line-Derived Neurotrophic Factor (GDNF) in MDD

Glial pathology in depressive disorder is well-documented by a number of quantitative studies on postmortem fronto-limbic brain regions. The density of astrocyte cell bodies immunoreactive for glial fibrillary acidic protein (GFAP) is consistently reduced in brain tissue specimens from depressed individuals, as is the expression of astrocyte proteins such as GFAP, GDNF, connexins, glutamate transporters, and glutamine synthetase [42]. Astrocytes play essential roles in maintaining brain homeostasis and neuronal functions, and also mediate innate immunity and inflammatory responses in the brain. GDNF is a member of the transforming growth factor β superfamily that was isolated and purified from the conditioned medium of cultured rat glial cells of the B49 cell line [43]. GDNF consists of 134 amino acids with near-identical sequence in rats and humans. The widespread distribution of GDNF and its receptors in various regions of the adult brain suggests a role in maintaining neuronal circuits in the mature central nervous system (CNS) [44]. An increase in astrocyte GDNF synthesis and protein expression is believed to play an active role in neuronal survival and plasticity after excitotoxic damage [45], while experimental strategies of GDNF delivery by astrocytes have shown neuroprotective effects in vivo for dopaminergic neurons [46].

Animal studies have revealed that GDNF affects cognitive function, including learning and memory [47], while GDNF infusion increases hippocampal neurogenesis [48]. $GDNF^{+/-}$ mutant mice show abnormal hippocampal synaptic transmission [49], and GDNF overexpression in astrocytes of the hippocampal CA1 region can improve spatial learning and memory performance in cognitively impaired aging rats by enhancing local cholinergic, dopaminergic, and serotonergic transmission [50].

Several different classes of antidepressant including amitriptyline, clomipramine, mianserin, fluoxetine, and paroxetine have been shown to increase *GDNF* mRNA expression and protein secretion in rat C6 glioblastoma cells, while amitriptyline increased transcript expression in rat astrocytes when administered at concentrations comparable to those used in clinical trials. The results indicate that these drugs act through modulation of GDNF to improve the function of both glia and neurons [44].

http://dx.doi.org/10.5772/59373

Some animal studies have reported that chronic treatment with several classes of antidepressant or mood stabilizer has no affect on *GDNF* mRNA and protein expression in various areas of the rat brain [51-52]. However, in Flinders Resistant Line rats, chronic lithium treatment increased GDNF protein in the frontal and occipital cortices, decreased GDNF in the hippocampus, but did not alter GDNF level in the striatum [53]. In a rat model of chronic unpredictable stress-induced depression, *GDNF* mRNA and protein levels were significantly decreased in the hippocampus; this was reversed by clomipramine treatment [54]. The neuroprotective effects of clomipramine in the hippocampus suggest that GDNF is a viable target for novel antidepressant drugs. Chronic stress increased DNA methylation and histone modification in the promoter region of the *GDNF* gene and altered the control of behavioral responses in animal models of depression, effects that were reversed by antidepressant treatment [55].

The reduction in the volume of the cortex and limbic system that is observed in depressed patients is primarily associated with a decrease in glial cell numbers and to a lesser degree with decreased neuronal density and size [42]. In postmortem brain tissue from a small number of subjects with recurrent depression, increased GDNF level was detected in the parietal cortex relative to controls, and it was postulated that a loss of neurons and glia leads to an upregulation of GDNF as a compensatory response, which has also been reported in brain injury and animal models [56].

Electroacupuncture (EA) stimulation has been used for several decades to treat various mood disorders. A recent meta-analysis of 20 clinical trials found that acupuncture monotherapy was as effective as antidepressant treatment in terms of treatment response and alleviating the severity of symptoms [57]. Serum GDNF was increased by treatment with EA or fluoxetine in depressed patients and was associated with decreased Hamilton Depression Rating Scale scores in depressed patients [58]. In rats with transection of the medial forebrain bundle, EA upregulated *GDNF* mRNA expression in the brain [59]; acupuncture was also found to stimulate GDNF expression in the brain of adult cats [60]. These results suggest that EA can improve psychological symptoms of depression by altering *GDNF* expression, thereby halting or slowing neurodegeneration.

As in the case of antidepressants, the therapeutic effects of electroconvulsive therapy (ECT) involve stimulation of proliferation in neural progenitors via upregulation of specific signal transduction pathways [61]. ECT stimulates glial cell proliferation in the prefrontal cortex of rats by causing a reduction in the expression of *Sprouty2*, an inhibitor of cell proliferation [62]. ECT was also shown to decrease GDNF concentration in the hippocampus and striatum of the adult rat brain [63], although it is unknown whether this is due to decreased synthesis or increased release of the protein. However, it underscores the finding that acute and chronic ECT enhanced the mRNA expression of the GDNF receptor in the dentate gyrus of the hippocampus in rats [51]. Our research has demonstrated that serum GDNF level rises following ECT in patients with drug-resistant depression.

Adult mice heterozygous for a null mutation at the *GDNF* locus exhibited a lower level of *GDNF* mRNA in all brain regions examined. These mice also showed significant behavioral deficits in the Morris water maze spatial learning paradigm [47]. Overexpression of a *GDNF* transgene in hippocampal astrocytes induced the recovery of spatial cognitive abilities in aged rats [50]. GDNF concentration was positively correlated with performance on the Wisconsin Card Sorting Test (WCST) conceptual level responses and negatively correlated with performance on the WCST preservative errors in depressed patients [64]. Moreover, increased plasma GDNF level was positively correlated with the Digit Span Test backward score and negatively associated with Trail Making Test B performance in late-onset depression patients [65]. These results indicate that increased GDNF may protect against neuronal damage and consequent cognitive impairment in depressed individuals.

A recent study evaluated the effect of 21 single nucleotide polymorphisms in the *GDNF* gene on the efficacy of paroxetine in patients with MDD, and found that the A allele for rs 2973049 and the T allele for rs 2216711 were correlated with paroxetine response and gender [66].

The neuroprotective effects of GDNF may be due in part to its ability to protect neurons from oxidative stress. Postmortem studies indicate that patients with recurrent depressive disorder have increased oxidative stress in some brain regions, such as the frontal cortex, thalamus, and putamen [67]. Subchronic infusion of recombinant human GDNF increased superoxide dismutase, catalase, and glutathione peroxidase activities in rat striatum, suggesting that GDNF may have antioxidant properties [68]. GDNF also protected human mesencephalic neuron-derived cells from oxidative injury [69].

We found that serum GDNF level was decreased in antidepressant-free patients with MDD and was not correlated with depression severity. We also found that decreased serum GDNF level in naive patients recovered to normal levels after treatment with antidepressants [70]. Reduced *GDNF* mRNA expression was detected in peripheral white blood cells of MDD patients in a depressive state but not in those in a remissive state, suggesting that the alteration in GDNF level is state-dependent [71]. Circulating serum GDNF level was also decreased in patients with late-in-life depression, and this was negatively correlated to the disease severity [72]. However, plasma GDNF level was higher in euthymic patients with bipolar disorder and in patients with late-onset depression [65,73]. This inconsistency may be due to confounding effects of age, gender, or concurrent physical illness. One study has shown that patients with bipolar and unipolar affective disorder in remission had a decreased GDNF level compared to unaffected controls, which was not correlated with antidepressant treatment [74].

4. Insulin-like Growth Factor (IGF) in MDD

There is increasing evidence that IGF-1 plays an important role in diseases affecting the CNS. IGF-1 increases the synthesis and activity of BDNF [75], and both are required for neuronal survival and synaptic plasticity in the brain [76]. IGF-1 also enhances proliferation, survival, differentiation, and maturation of all CNS cells and has demonstrated neurotrophic, neurogenic, and neuroprotective functions [77]. IGF-1 is the only neurotrophic factor known to be regulated by the immune system, the dysregulation of which is implicated in the pathogenesis of depression.

http://dx.doi.org/10.5772/59373

The *Igf1* gene in humans is located on chromosome 12q22–23. IGF-1 is a small (7.5-kDa) polypeptide that, along with IGF-2, insulin, their respective receptors, and six IGF-binding proteins, constitute the insulin-like growth factor family [78]. IGF-1 and its receptor are widely distributed in all cell types of the adult brain. Although it can penetrate the blood-brain barrier, IGF-1 is also produced by various cells in the CNS and peripheral nervous system [79]. *Igf1* mRNA expression in the CNS is low during early organogenesis, but increases at later developmental stages [79]; after the brain is formed, low levels of expression are restricted to a few regions. However, the adult brain also receives IGF-1 from the serum where the peptide is abundant. Additionally, IGF-1 expression remains high in the adult brain, especially in areas with large projection neurons such as the cerebellum, olfactory bulb, hypothalamus, hippocampus, cortex, and retina [80].

Individuals with depression have reduced hippocampal volume as compared to controls. Post mortem and brain imaging studies have revealed atrophy in the hippocampus of depressed patients, which may be reversed by antidepressant treatment [23]. Peripheral infusion of IGF-1 induced neurogenesis in adult rat hippocampus, and IGF-1 stimulates proliferation in adult rat hippocampal progenitor cells [81], while in vitro studies found that IGF-1 increased the total number of progenitor cells and promoted the specification of a neuronal lineage from precursors [79]. Transgenic mice lacking IGF-1 or IGF-1R show severe delays in brain development [82]; mice overexpressing IGF-1 in the brain exhibit an increased numbers of neurons and synapses in the dentate gyrus, while IGF-1 knockout mice have a decreased number of granule cells in this region [83-84].

In animal models of depression, the specificity of the models used, the immune status of the tested animals, or animals' age and gender can influence the levels of growth factors including that of IGF-1. In one study using the Cre-loxP system to knock out the *Igf1* gene in the liver, hippocampal CA1 neurons, or both, IGF-1 deficiency at adulthood was sufficient to induce a depressive phenotype in mice, suggesting that low brain IGF-1 level heightens the risk for depression; moreover, these effects are not ameliorated by increased local IGF-1 production or transport [85]. A prenatal stress model of depression found no changes in peripheral IGF-1 levels between stressed and control rats, but a significant decrease was observed in the hippocampus and frontal cortex [79].

Early adverse experiences contribute to the development of vulnerability to stress and increase the risk of stress-related psychiatric disorders in adulthood. For instance, maternal separation during critical periods of brain development can lead to learning disabilities, behavioral anomalies, or psychiatric disorders in later life [86]. Repeated maternal separation of neonatal rats caused prolonged and abnormal fluctuations in the expression of BDNF and IGF-1 and their respective receptors TrkB and IGF-1R in the cerebral cortex [86]. Another study found that maternal separation alone or in combination with a single episode of restraint stress decreased the mRNA expression of *IGF-1R* and *IGF binding protein-2* in the adult rat hippocampus. Given that the activation of IGF signaling plays a role in development and neuroprotection of the CNS, this downregulation of IGF signaling likely contributes to the development of stress vulnerability in adulthood, which can in turn precipitate the onset of depression [87].

The potential antidepressant activity of IGF-1 has been examined in various animal models of depression in which behavioral tests such as forced swimming and tail suspension were used to evaluate antidepressant effects. These studies have consistently shown that IGF-1 treatment has antidepressant-like effects and normalizes behavioral disturbances in depressive animals [88-90]. In addition, repeated administration of fluoxetine induced the upregulation of IGF-1 and its receptor in the frontal cortex but a downregulation in the hippocampus [91]. IGF-1 level was also upregulated in the adult rat hippocampus after chronic administration of venlafaxine [92]

Increased IGF-1 level in the blood of depressed patients has been observed in clinical studies [93-94], while others have reported decreased peripheral IGF-1 concentration in patients, possibly due to overactivation of the hypothalamic-pituitary-adrenal axis. A recent study demonstrated that a low IGF-1 level in females and a high level in males can predict the incidence of depressive disorder 5 years later [95]. IGF-1 concentration was also found to decline during antidepressant treatment in patients, albeit only in responders [93].

5. Vascular Growth Factor (VGF) in MDD

VGF, originally cloned as a target of NGF regulation in PC12 cells, is also induced by BDNF in cortical and hippocampal neurons in vitro and in vivo [96]. The *VGF* gene has been highly conserved throughout evolution and is located on chromosome 7q22 in humans and chromosome 5 in mice. VGF contains a cyclic AMP response element-binding protein (CREB)-binding site within its promoter that is critical for BDNF-induced *VGF* expression; thus, CREB is a factor regulating both *BDNF* and *VGF* expression [97]. VGF also contributes to synaptic plasticity by inducing the secretion of BDNF or other neuromodulatory peptides in a positive feedback loop, and mediates the long-term effects of BDNF-induced synaptic strengthening through its activity at excitatory synapses [96]. VGF is also involved in regulating energy balance via hypothalamic and autonomic outflow pathways that govern peripheral energy expenditure [98]. VGF is widely expressed in neurons and is detected in the olfactory system and several areas of the brain, including the cerebral cortex, hypothalamus, and hippocampus [99].

$VGF^{+/-}$ mice have no gross abnormalities in brain morphology and exhibit normal anxiety levels. However, these mice show neurological and behavioral deficits akin to depression [100-102], suggesting that a reduced VGF level may account for depression in humans. Chronic treatment with different classes of antidepressant such as imipramine, fluoxetine, and duloxetine has been shown to modulate VGF expression [100-101,103], while infusion of VGF into the midbrain or hippocampus produces antidepressant effects in the learned helplessness paradigm of depression, as well as in the tail suspension and forced swim tests, which are used to evaluate the action of antidepressants [101-102]. In one of these studies, exercise increased VGF in the hippocampus of wild-type mice and induced an antidepressant-like response in the forced swim test; the increase was less pronounced in $VGF^{+/-}$ mice, which also failed to show behavioral improvement resulting from exercise in the forced swim test. Moreover, the administration of VGF peptide induced a robust, dose-dependent antidepressant-like response in the forced swim and tail suspension tests. In another study, VGF-derived peptide protected primary cultures of rat cerebellar granule cells from serum and potassium deprivation-induced cell death in a dose-and time-dependent manner [104].

Chronic antidepressant treatment in adult rodents increases neurogenesis in the dentate gyrus of the hippocampus, which may be a common mechanism by which antidepressants induce their therapeutic effects [105]. VGF peptide similarly enhanced neurogenesis of hippocampal cells and may favor the differentiation of proliferating progenitors into neurons over glia [101]. Although the precise relationship between depression and neurogenesis remains to be elucidated, stimulating cell proliferation may be one way in which VGF exerts antidepressant effects.

ECT is a highly effective and rapid treatment for depressed patients who do not respond to antidepressants. While the molecular mechanisms underlying the therapeutic efficacy of ECT are not fully understood, it is thought that ECT and antidepressants increase the expression of select neurotrophic factors that reverse or block the atrophy and cell loss resulting from stress and depression. ECT was shown to increase the level of VGF in the hippocampus of rats [103,106], and a decrease in *VGF* mRNA level was observed in drug-free depressed patients with respect to controls, although there was no correlation between *VGF* mRNA levels and the severity of the illness. Interestingly, 12 weeks of treatment with escitalopram increased VGF expression albeit only in responder patients, suggesting that changes in the expression of the neuropeptide may explain the mechanism of the drug response [107].

In the European Genome-based Therapeutic Drugs for Depression study, nine psychiatric centers in eight European countries recruited 811 adult outpatients suffering from unipolar depression of at least moderate severity. These patients showed significant upregulation in leukocyte *VGF* mRNA expression level after 8 weeks of treatment in both responder and non-responder patients [108].

Findings from human postmortem studies of bipolar patients indicate that VGF is downregulated in the CA regions of the hippocampus and Brodmann's area 9 of the dorsolateral prefrontal cortex [109]. Importantly, VGF showed effects similar to those of lithium, which is used to treat bipolar disorder. However, another study found an increase in VGF23–62 peptide level in the cerebrospinal fluid of 16 patients diagnosed with MDD [110]. The association between *VGF* gene polymorphisms and depression or bipolar disorder has not been examined thus far.

6. Fibroblast Growth Factor (FGF) in MDD

FGFs and their receptors constitute an elaborate signaling system that is organized in dynamic spatial and temporal expression patterns. There are at least 23 members of the FGF family, 22 of which are distributed throughout the CNS in humans along with five FGF receptors. FGF1 and FGF2, which lack a signal sequence, are secreted and directly regulate intracellular signaling cascades in target cells. FGF ligands share a conserved central domain of about 120 amino acids that binds heparin and is required for stable FGF ligand-receptor interactions [111]. FGFRs are tyrosine kinase receptors with three immunoglobulin-like domains (D1, D2, and D3) [112].

FGF signaling has been implicated in a variety of mood disorders, including MDD. The neurotrophic hypothesis of depression posits that neurogenesis and neuronal plasticity are affected by the imbalance in growth factor levels. FGF2 attenuates the reduction in hippocampal volume and promotes hippocampal neurogenesis after traumatic brain injury in mice [113]. Post-mortem examinations have shown that the expression of FGF1, FGF2, and receptors FGFR2 and FGFR3 is decreased in the frontal cortical area in MDD relative to controls. FGF signaling is also upregulated by treatment with antidepressants [114]. A lower serum FGF-2 level has been reported in MDD patients, which may be reversed by antidepressants [115]. In an animal model, a single subcutaneous injection of FGF2 administered to rats on postnatal day 1 increased cell survival in the DG 3 weeks later, producing a larger hippocampus with more cells [116]; a similar effect was observed in the adult brain [112].

FGFR1 conditional knockout mice lacking *FGFR1* expression in the telencephalon at mid-neurogenesis showed that the *FGFR1* gene is required during hippocampal development [111] for the proliferation of hippocampal progenitor and stem cells [117]. Loss of FGFR1 function results in decreased proliferation of neural progenitor cells in the hippocampus and depletion of the FGF2-sensitive hippocampal neural stem cell pool, which leads to permanent atrophy of this brain area. $FGF2^{-/-}$ mice showed reduced numbers of cortical neurons by the end of neurogenesis, especially of large neurons in deeper cortical layers of the frontal cerebral cortex [118]. Progenitors of hippocampal granule cells in the DG in $FGF2^{-/-}$ mice showed reduced proliferation in response to kainic acid injection or middle cerebral artery occlusion, which was reversed by FGF2 injection [119]. Thus, FGF2 induces neurogenesis, alters neuronal morphology, and modulates gene expression in the hippocampus.

FGF2 has antidepressant-like effects when administered later in adulthood. FGF2 infusions had both antidepressant and anxiolytic effects in behavioral models of depression and anxiety [120]. FGF activity has also been linked to the response to antidepressant medications in humans, and some studies have found a correlation between *FGF* transcript expression and major depression.

Affymetrix microarray analyses of cortical brain regions detected a significant number of *FGF*-related ligand and receptor genes that are differentially expressed in depressed individuals, including *FGF1, FGF2, FGFR2,* and *FGFR3,* which are downregulated, and *FGF9* and *FGF12,* which are upregulated [114]. Depressed subjects receiving selective serotonin reuptake inhibitor (SSRI) treatment showed increased FGFR expression as compared to those who were not using these drugs. One recent study also demonstrated alterations in *FGF9* and *FGFR3* transcript expression in the locus coeruleus (LC) of individuals with MDD [121]. Serum FGF2 level in MDD patients was significantly lower than that of healthy controls [115], and in postmortem samples of MDD patients, single nucleotide polymorphisms in *FGF2* (rs1048201, rs1449683, and rs308393) were correlated with side effects and differential treatment response to antidepressants [122].

7. Neurotropin-3 (NT3) in MDD

NT3 is a member of the neurotrophin family of proteins that supports the survival of specific types of neuron [123]. NT3 is distributed throughout the DG and promotes hippocampal plasticity by regulating neurogenesis via binding of TrkB and TrkC tyrosine kinase neurotrophin receptors [124].

NT3 has been implicated in the pathogenesis of depression and the therapeutic mechanism of antidepressants. Hippocampal atrophy has been consistently demonstrated as one of the predominant pathophysiological changes in subjects with a history of MDD, and is correlated with duration of the illness [125]. NT3 prevented the degeneration of noradrenergic neurons in the LC [126] that are associated with the pathophysiology of major depression [127]. NT3 infusions cause an upregulation in the mRNA level of *BDNF* in the cerebral cortex [128] and produces BDNF-like effects including the phosphorylation of cortical tyrosine kinase B [123]. NT3 may be involved in the modulation of BDNF signaling in differentiating hippocampal neurons [129], and was shown to modulate monoamine neurotransmitters such as 5-hydroxytryptophan and noradrenaline [130] and activate noradrenergic neurons in the LC [131]. Chronic stress causes structural changes and neuronal damage in the brain, especially in the hippocampus, which can lead to the development of depression [132]. *NT3* mRNA levels were upregulated in the DG and hippocampus in response to repeated immobilization stress in rodents [133]. NT3 is expressed throughout the hippocampus and promotes plasticity by regulating neurogenesis. A 500-ng dose of NT3 injected into the cerebral ventricle of mice at the onset of the dark period enhanced the time spent in non-rapid eye movement sleep [134]. Lithium and valproate, which are mood-stabilizing agents, increased hippocampal NT3 levels in an animal model of mania [135], highlighting a role for NT3 in mood disorders besides depression.

NT3 antagonizes the proliferative effects of basic FGF, and enhances neuronal differentiation, while blocking NT3 function leads to a decrease in neurogenesis [136]. In NT-3 mutant mice with brain-specific *NT3* deletion, the differentiation of the neuronal precursor cells was impaired in the DG, resulting in a decrease in the production of differentiated neurons [124].

Postmortem analyses have demonstrated altered levels of neurotrophic factor expression in the brains of patients with mood disorder. Serum NT3 levels were increased during acute mood states of bipolar depression patients as compared to euthymic patients and normal controls [137]. In another study, serum NT3 level in drug-free and medicated patients with bipolar disorder during manic and depressive episodes was increased relative to controls, while there was no difference between medicated and drug-free patients [138]. *NT3* mRNA was downregulated in peripheral white blood cells of patients with MDD and bipolar disorder during depressed and euthymic states, but not those in remission, suggesting that reduced *NT3* expression is state-dependent and associated with the pathophysiology of major depression [71]. However, in elderly MDD patients, NT3 levels in the cerebrospinal fluid were significantly elevated as compared to patients with Alzheimer's disease or mentally healthy controls [139]. The reason for these contradictory findings is unknown, but could indicate that there are other age-dependent factors that modulate the effects of NT3.

8. Nerve Growth Factor (NGF) in MDD

NGF was the first identified neurotrophin in a family of structurally similar growth factors. In the CNS, NGF is involved in neuronal survival, protection of sympathetic and cholinergic neurons against neurodegeneration, and in the modulation of the immune response as well as learning and memory [140].

NGF has demonstrated neuroprotective effects on basal forebrain cholinergic neurons in Alzheimer's disease patients [141]. However, the injection of NGF antibody was shown to induce the death of sympathetic neurons in mouse, rat, cat, and rabbit models [142]. Mouse models of anxiety, chronic stress, and depression involving learned helplessness, threatening or painful stimuli, maternal deprivation, and other factors induce a reduction in NGF in the frontal cortex, amygdala, hippocampus, and nucleus accumbens [143-145]. In MDD patients, serum NGF level is higher than in controls, and has been associated with the severity of depressive symptoms in women [146].

Whether alterations in NGF are state-or trait-dependent is under debate. In mood disorders, a state-related phenomenon appears and then disappears with mood state; in contrast, a trait-related phenomenon occurs regardless of variations in the clinical state. One study of in-and outpatients with mood disorders including uni-and bipolar depression and bipolar mania found no differences or changes in NGF levels in patients with depressive episodes and after 8 weeks of medical treatment [147], consistent with previous findings that NGF levels do not vary among depressed patients [148-150].

Nonetheless, several clinical studies have reported that serum NGF concentration is decreased relative to healthy controls in MDD patients [151-152], including those receiving duloxetine [153] and those experiencing depression late in life [154]. In contrast, other studies have reported increased NGF in patients with elevated levels of stress and severe depression [155]. The severity of washing symptoms is correlated with an upregulation in NGF in patients with obsessive-compulsive disorder [156], while another study found a positive correlation between plasma NGF level and disease duration in patients with bipolar mood disorders [157].

Various studies have investigated the effects of ECT on NGF levels in animals and humans. Electroconvulsive stimuli administered once daily for 8 days increased NGF level in the frontal cortex of adult rats [63]. Repeated exposure to electroconvulsive stimuli also increased TrkA and NGF protein levels in the rat hippocampus [158], suggesting that NGF may play a role in the mechanism of action of electroconvulsive treatment. In contrast to animal studies, ECT treatment has not been found to affect NGF levels in human patients. In patients with treatment-resistant major depression [159] or bipolar disorder with depression [160], NGF levels were not significantly increased by ECT, even with concurrent administration of antidepressants and psychotherapy.

9. Conclusion

Several preclinical and clinical observations indicate that depression may be associated with the inability of neural systems to exhibit adaptive plasticity. Given the role of neurotrophic factors in neural and structural plasticity, and that depression and antidepressants exert opposite actions on neurotrophic factors expression and functions, it is apparent that neurotrophic factors signaling may be crucial in the pathophysiology of depression and in the mechanism of action of antidepressants. However, future work will be necessary to determine whether neurotrophic factors is a risk factor for initiation or maintenance or in the recovery process with respect to MDD and how its circuit-level function contributes at MDD stages. In addition, the search for more effective and applicable neurotrophic factors-based therapies is crucial.

Acknowledgements

This chapter was partially supported by Jiangsu Province Natural Science Foundation (BK2011434) and Yangzhou Municipal Key Technology Problems Foundation (YZ2010089).

Author details

Xiaobin Zhang*, Jin Li, Weiwei Sha and Ru Bu

*Address all correspondence to: zhangxiaobim@163.com

Department of Psychiatry, Affiliated WuTaiShan Hospital of Medical College of Yangzhou University, Yangzhou, China

References

[1] Belmaker RH, Agam G. Major depressive disorder. The New England Journal of Medicine 2008;358(1) 55-68.

[2] Huang EJ, Reichardt LF. Neurotrophins: roles in neuronal development and function. Annual Review of Neuroscience 2001;24 677-736.

[3] Lessmann V, Gottmann K, Malcangio M. Neurotrophin secretion: current facts and future prospects. Progress in Neurobiology 2003;69(5) 341-374.

[4] Reichardt L.F. Neurotrophin-regulated signalling pathways. Philosophical Transactions of the Royal Society of London. Series B, Biological Sciences 2006;361(1473) 1545-1564.

[5] Dwivedi Y. Brain-derived neurotrophic factor: role in depression and suicide. Neuropsychiatric disease and Treatment 2009;5 433-449.

[6] Siuciak JA, Lewis DR, Wiegand SJ, Lindsay RM. Antidepressant-like effect of brain-derived neurotrophic factor (BDNF). Pharmacology, Biochemistry, and Behavior 1997;56(1) 131-137.

[7] Shirayama Y, Chen ACH, Nakagawa S, Russell DS, Duman RS. Brain-derived neurotrophic factor produces antidepressant effects in behavioral models of depression. The Journal of neuroscience: the Official Journal of the Society for Neuroscience 2002;22(8) 3251-3261.

[8] Jiang C, Salton SR. The Role of Neurotrophins in Major Depressive Disorder. Translational Neuroscience 2013;4(1) 46-58.

[9] Allaman I, Papp M, Kraftsik R, Fiumelli H, Magistretti PJ, Martin JL. Expression of brain-derived neurotrophic factor is not modulated by chronic mild stress in the rat hippocampus and amygdala. Pharmacological reports: PR 2008;60(6) 1001-1007.

[10] Lucca G, Comim CM, Valvassori SS, Pereira JG, Stertz L, Gavioli EC, Kapczinski F, Quevedo J. Chronic mild stress paradigm reduces sweet food intake in rats without affecting brain derived neurotrophic factor protein levels. Current Neurovascular Research. 2008;5(4) 207-213.

[11] Marmigère F, Givalois L, Rage F, Arancibia S, Tapia-Arancibia L. Rapid induction of BDNF expression in the hippocampus during immobilization stress challenge in adult rats. Hippocampus 2003;13(5) 646-655.

[12] Uysal N, Sisman AR, Dayi A, Ozbal S, Cetin F, Baykara B, Aksu I, Tas A, Cavus SA, Gonenc-Arda S, Buyuk E. Acute footshock-stress increases spatial learning-memory and correlates to increased hippocampal BDNF and VEGF and cell numbers in adolescent male and female rats. Neuroscience Letters 2012;514(2) 141-146.

[13] Castrén E, Võikar V, Rantamäki T. Role of neurotrophic factors in depression. Current Opinion in Pharmacology 2007;7(1) 18-21.

[14] Saarelainen T, Hendolin P, Lucas G, Koponen E, Sairanen M, MacDonald E, Agerman K, Haapasalo A, Nawa H, Aloyz R, Ernfors P, Castrén E. Activation of the TrkB neurotrophin receptor is induced by antidepressant drugs and is required for antidepressant induced behavioral effects. The Journal of Neuroscience: the Official Journal of the Society for Neurosciencei 2003;23(1) 349-357

[15] Nibuya M, Morinobu S, Duman RS. Regulation of BDNF and trkB mRNA in rat brain by chronic electroconvulsive seizure and antidepressant drug treatments. The Jour-

nal of neuroscience: the Official Journal of the Society for Neuroscience 1995;15(11) 7539-7547.

[16] Müller MB, Toschi N, Kresse AE, Post A, Keck ME. Long term repetitive transcranial magnetic stimulation increases the expression of brain-derived neurotrophic factor and cholecystokinin mRNA, but not neuropeptide tyrosine mRNA in specific areas of rat brain. Neuropsychopharmacology: Official Publication of the American College of Neuropsychopharmacology. 2000;23(2) 205-215.

[17] Castrén E, Rantamäki T. The role of BDNF and its receptors in depression and antidepressant drug action: Reactivation of developmental plasticity. Developmental Neurobiology 2010;70(5) 289-297.

[18] Yu H, Chen ZY. The role of BDNF in depression on the basis of its location in the neural circuitry. Acta Pharmacologica Sinica 2011;32(1) 3-11.

[19] Castrén E. Neurotrophic effects of antidepressant drugs. Current Opinion in Pharmacology 2004;4(1) 58-64.

[20] Chen B, Dowlatshahi D, MacQueen GM, Wang JF, Young LT. Increased hippocampal BDNF immunoreactivity in subjects treated with antidepressant medication. Biological Psychiatry 2001;50(4) 260-265.

[21] Chen ZY, Patel PD, Sant G, Meng CX, Teng KK, Hempstead BL, Lee FS. Variant brain-derived neurotrophic factor (BDNF) (Met66) alters the intracellular trafficking and activity-dependent secretion of wild-type BDNF in neurosecretory cells and cortical neurons. The Journal of Neuroscience: the Official Journal of the Society for Neuroscience 2004;24(18) 4401-4411.

[22] Egan MF, Kojima M, Callicott JH, Goldberg TE, Kolachana BS, Bertolino A, Zaitsev E, Gold B, Goldman D, Dean M, Lu B, Weinberger DR. The BDNF val66met polymorphism affects activity-dependent secretion of BDNF and human memory and hippocampal function. Cell 2003;112(2) 257-269.

[23] Masi G, Brovedani P. The hippocampus, neurotrophic factors and depression: possible implications for the pharmacotherapy of depression. CNS Drugs 2011;25(11) 913-931.

[24] Frodl T, Schüle C, Schmitt G, Born C, Baghai T, Zill P, Bottlender R, Rupprecht R, Bondy B, Reiser M, Möller HJ, Meisenzahl EM. Association of the brain derived neurotrophic factor Val66Met polymorphism with reduced hippocampal volumes in depression. Archives of General Psychiatry 2007;64(4) 410-416.

[25] Hong CJ, Huo SJ, Yen FC, Tung CL, Pan GM, Tsai SJ. Association study of a brain-derived neurotrophic-factor genetic polymorphism and mood disorders, age of onset and suicidal behavior. Neuropsychobiology 2003;48(4) 186-189.

[26] Tsai SJ, Cheng CY, Yu YW, Chen TJ, Hong CJ. Association study of a brain-derived neurotrophic-factor genetic polymorphism and major depressive disorders, sympto-

matology, and antidepressant response. American Journal of Medical Genetics. Part B, Neuropsychiatric Genetics : the Official Publication of the International Society of Psychiatric Genetics 2003;123B(1) 19-22.

[27] Hwang JP, Tsai SJ, Hong CJ, Yang CH, Lirng JF, Yang YM. The Val66Met polymorphism of the brain-derived neurotrophic-factor gene is associated with geriatric depression. Neurobiology of Aging 2006;27(12) 1834-1837.

[28] Zou YF, Ye DQ, Feng XL, Su H, Pan FM, Liao FF. Meta-analysis of BDNF Val66Met polymorphism association with treatment response in patients with major depressive disorder. European neuropsychopharmacology: the Journal of the European College of Neuropsychopharmacology 2010;20(8) 535-544.

[29] Kato M, Serretti A. Review and meta-analysis of antidepressant pharmacogenetic findings in major depressive disorder. Molecular Psychiatry 2010;15(5) 473-500.

[30] Domschke K, Lawford B, Laje G, Berger K, Young R, Morris P, Deckert J, Arolt V, McMahon FJ, Baune BT. Brain-derived neurotrophic factor (BDNF) gene: no major impact on antidepressant treatment response. The International Journal of Neuropsychopharmacology /Official Scientific Journal of the Collegium Internationale Neuropsychopharmacologicum (CINP) 2010;13(1) 93-101.

[31] Verhagen M, van der Meij A, van Deurzen PA, Janzing JG, Arias-Vásquez A, Buitelaar JK, Franke B. Meta-analysis of the BDNF Val66Met polymorphism in major depressive disorder: effects of gender and ethnicity. Molecular Psychiatry 2010;15(3) 260-271.

[32] Dwivedi Y, Rizavi HS, Conley RR, Roberts RC, Tamminga CA, Pandey GN. Altered gene expression of brain-derived neurotrophic factor and receptor tyrosine kinase B in postmortem brain of suicide subjects. Archives of General Psychiatry 2003;60(8) 804-815.

[33] Karege F, Vaudan G, Schwald M, Perroud N, La Harpe R. Neurotrophin levels in postmortem brains of suicide victims and the effects of antemortem diagnosis and psychotropic drugs. Brain Research. Molecular Brain Research 2005;136 (1-2) 29-37.

[34] Karege F, Schwald M, Cisse M. Postnatal developmental profile of brain-derived neurotrophic factor in rat brain platelets. Neuroscience Letters 2002;328(3) 261-264.

[35] Karege F, Perret G, Bondolfi G, Schwald M, Bertschy G, Aubry JM. Decreased serum brain-derived neurotrophic factor levels in major depressed patients. Psychiatry Research. 2002;109(2) 143-148.

[36] Huang TL, Lee CT, Liu YL. Serum brain-derived neurotrophic factor levels in patients with major depression: effects of antidepressants. Journal of Psychiatric Research 2008;42(7) 521-525.

[37] Gonul AS, Akdeniz F, Taneli F, Donat O, Eker C, Vahip S. Effect of treatment on serum brain-derived neurotrophic factor levels in depressed patients. European Archives of Psychiatry and Clinical Neuroscience 2005;255(6) 381-386.

[38] Shimizu E, Hashimoto K, Okamura N, Koike K, Komatsu N, Kumakiri C, Nakazato M, Watanabe H, Shinoda N, Okada S, Iyo M. Alterations of serum levels of brain-derived neurotrophic factor (BDNF) in depressed patients with or without antidepressants. Biological Psychiatry 2003;54(1) 70-75.

[39] Lee BH, Kim H, Park SH, Kim YK. Decreased plasma BDNF level in depressive patients. Journal of Affective Disorders 2007;101(1-3) 239-244.

[40] Zhang X, Wang X, Sha W, Zhou H, Zhang Y. Val66Met polymorphism and serum brain-derived neurotrophic factor concentration in depressed patients. Acta Neuropsychiatrica 2011;23(5) 229-234

[41] Bus BA, Molendijk ML, Penninx BJ, Buitelaar JK, Kenis G, Prickaerts J, Elzinga BM, Voshaar RC. Determinants of serum brain-derived neurotrophic factor. Psychoneuroendocrinology 2011;36(2) 228-239.

[42] Rajkowska G, Stockmeier CA. Astrocyte pathology in major depressive disorder: insights from human postmortem brain tissue. Current Drug Targets 2013;14(11) 1225-1236.

[43] Lin LF, Doherty DH, Lile JD, Bektesh S, Collins F. GDNF: a glial cell line-derived neurotrophic factor for midbrain dopaminergic neurons. Science 1993;260(5111) 1130-1132.

[44] Hisaoka K, Nishida A, Koda T, Miyata M, Zensho H, Morinobu S, Ohta M, Yamawaki S. Antidepressant drug treatments induce glial cell line-derived neurotrophic factor (GDNF) synthesis and release in rat C6 glioblastoma cells. Journal of Neurochemistry 2001;79(1) 25-34.

[45] Marco S, Canudas AM, Canals JM, Gavald`a N, P′erez-Navarro E, Alberch J. Excitatory amino acids differentially regulate the expression of GDNF, neurturin, and their receptors in the adult rat striatum. Experimental Neurology 2002;174 (2) 243-252.

[46] Ericson C, Georgievska B, Lundberg C. Ex vivo gene delivery of GDNF using primary astrocytes transduced with a lentiviral vector provides neuroprotection in a rat model of Parkinson's disease. The European Journal of Neuroscience 2005;22 (11) 2755-2764.

[47] Gerlai R, McNamara A, Choi-Lundberg DL, Armanini M, Ross J, Powell-Braxton L, Phillips HS. Impaired water maze learning performance without altered dopaminergic function in mice heterozygous for the GDNF mutation. The European Journal of Neuroscience 2001;14(7) 1153-1163.

[48] Reyes JH, O'Shea KS, Wys NL, Velkey JM, Prieskorn DM, Wesolowski K, Miller JM, Altschuler RA. Glutamatergic neuronal differentiation of mouse embryonic stem

cells after transient expression of neurogenin 1 and treatment with BDNF and GDNF: in vitro and in vivo studies. The Journal of Neuroscience: the Official Journal of the Society for Neuroscience 2008;28(48) 12622-12631.

[49] Nanobashvili A, Airaksinen MS, Kokaia M, Rossi J, Asztély F, Olofsdotter K, Mohapel P, Saarma M, Lindvall O, Kokaia Z. Development and persistence of kindling epilepsy are impaired in mice lacking glial cell line-derived neurotrophic factor family receptor alpha 2. Proceedings of the National Academy of Sciences of the United States of America 2000;97(22) 12312-12317.

[50] Pertusa M, García-Matas S, Mammeri H, Adell A, Rodrigo T, Mallet J, Cristòfol R, Sarkis C, Sanfeliu C. Expression of GDNF transgene in astrocytes improves cognitive deficits in aged rats. Neurobiology of Aging 2008; 29(9) 1366-1379.

[51] Chen AC, Eisch AJ, Sakai N, Takahashi M, Nestler EJ, Duman RS. Regulation of GFRalpha-1 and GFRalpha-2 mRNAs in rat brain by electroconvulsive seizure. Synapse 2001;39(1) 42-50.

[52] Fukumoto T,Morinobu S, OkamotoY,Kagaya A,Yamawaki S. Chronic lithium treatment increases the expression of brain-derived neurotrophic factor in the rat brain. Psychopharmacology 2001;158(1) 100-106.

[53] Angelucci F, Aloe L, Jiménez-Vasquez P, Mathé AA. Lithium treatment alters brain concentrations of nerve growth factor, brain-derived neurotrophic factor and glial cell line-derived neurotrophic factor in a rat model of depression. The International Journal of Neuropsychopharmacology 2003;6(3) 225-231.

[54] Liu Q, Zhu HY, Li B, Wang YQ, Yu J, Wu GC. Chronic clomipramine treatment restores hippocampal expression of glial cell line-derived neurotrophic factor in a rat model of depression. Journal of Affective Disorders 2012;141(2-3) 367-372.

[55] Uchida S, Hara K, Kobayashi A, Otsuki K, Yamagata H, Hobara T, Suzuki T, Miyata N, Watanabe Y. Epigenetic status of Gdnf in the ventral striatum determines susceptibility and adaptation to daily stressful events. Neuron 2011;69(2) 359-372.

[56] Michel TM, Frangou S, Camara S, Thiemeyer D, Jecel J, Tatschner T, Zoechling R, Grünblatt E. Altered glial cell line-derived neurotrophic factor (GDNF) concentrations in the brain of patients with depressive disorder: a comparative post-mortem study. European Psychiatry: the Journal of the Association of European Psychiatrists 2008;23(6) 413-420.

[57] Zhang ZJ, Chen HY, Yip KC, Ng R, Wong VT. The effectiveness and safety of acupuncture therapy in depressive disorders: Systematic review and meta-analysis. Journal of Affective Disorders 2010;124 (1-2) 9-21.

[58] Sun H, Zhao H, Ma C, Bao F, Zhang J, Wang DH, Zhang YX, He W. Effects of electroacupuncture on depression and the production of glial cell line-derived neurotrophic factor compared with fluoxetine: a randomized controlled pilot study. The Journal of

Alternative and Complementary Medicine: Research on Paradigm, Practice, and Policy 2013;19(9) 733-739.

[59] Liang XB, Luo Y, Liu XY, Lu J, Li FQ, Wang Q, Wang XM, Han JS. Electro-acupuncture improves behavior and upregulates GDNF mRNA in MFB transected rats. Neuroreport 2003;14(8):1177-1181.

[60] Wang TT, Yuan Y, Kang Y, Yuan WL, Zhang HT, Wu LY, Feng ZT. Effects of acupuncture on the expression of glial cell line-derived neurotrophic factor (GDNF) and basic fibroblast growth factor (FGF-2/bFGF) in the left sixth lumbar dorsal root ganglion following removal of adjacent dorsal root ganglia. Neuroscience Letters 2005;382(3) 236-241.

[61] Coyle JT, Duman RS. Finding the intracellular signaling pathways affected by mood disorder treatments. Neuron 2003;38(2) 157-160.

[62] Ongür D, Pohlman J, Dow AL, Eisch AJ, Edwin F, Heckers S, Cohen BM, Patel TB, Carlezon WA Jr. Electroconvulsive seizures stimulate glial proliferation and reduce expression of Sprouty2 within the prefrontal cortex of rats. Biological Psychiatry 2007;62(5) 505-512.

[63] Angelucci F, Aloe L, Jiménez-Vasquez P, Mathé AA. Electroconvulsive stimuli alter the regional concentrations of nerve growth factor, brain-derived neurotrophic factor, and glial cell line-derived neurotrophic factor in adult rat brain. The Journal of ECT 2002;18(3) 138-143.

[64] Zhang X, Ru B, Sha W, Xin W, Zhou H, Zhang Y. Performance on the Wisconsin card-sorting test and serum levels of glial cell line-derived neurotrophic factor in patients with major depressive disorder. Asia-Pacific psychiatry: Official Journal of the Pacific Rim College of Psychiatrists 2014.6 302-307.

[65] Wang X, Hou Z, Yuan Y, Hou G, Liu Y, Li H, Zhang Z. Association study between plasma GDNF and cognitive function in late-onset depression. Journal of Affective Disorders 2011;132(3) 418-421.

[66] Wang XC, Xu DJ, Chen GH, Xia Q, Liu LN. Association of Two Neurotrophic Factor Polymorphisms with Efficacy of Paroxetine in Patients with Major Depressive Disorder in a Chinese Population. Therapeutic Drug Monitoring, 2014 Feb 26. [Epub ahead of print]

[67] Michel TM, Frangou S, Thiemeyer D, Camara S, Jecel J, Nara K, Brunklaus A, Zoechling R, Riederer P. Evidence for oxidative stress in the frontal cortex in patients with recurrent depressive disorder-a postmortem study. Psychiatry Research 2007;151 (1-2) 145-150.

[68] Chao CC, Lee EH. Neuroprotective mechanism of glia cell line-derived neurotrophic factor on dopamine neurons: role of antioxidation. Neuropharmacology 1999;38(6) 913-916.

[69] Ortiz-Ortiz MA, Morán JM, Ruiz-Mesa LM, Bonmatty RG, Fuentes JM. Protective effect of the glial cell line-derived neurotrophic factor (GDNF) on human mesencephalic neuron-derived cells against neurotoxicity induced by paraquat. Environmental Toxicology and Pharmacology 2011;31(1) 129-136.

[70] Zhang X, Zhang Z, Xie C, Xi G, Zhou H, Zhang Y, Sha W. Effect of treatment on serum glial cell line-derived neurotrophic factor in depressed patients. Progress in Neuro-psychopharmacology & Biological Psychiatry 2008;32(3) 886-890.

[71] Otsuki K, Uchida S, Watanuki T, Wakabayashi Y, Fujimoto M, Matsubara T, Funato H, Watanabe Y. Altered expression of neurotrophic factors in patients with major depression. Journal of Psychiatric Research 2008;42(14) 1145-1153.

[72] Diniz BS, Teixeira AL, Miranda AS, Talib LL, Gattaz WF, Forlenza OV. Circulating Glial-derived neurotrophic factor is reduced in late-life depression. Journal of Psychiatric Research 2012;46 (1) 135-139.

[73] Barbosa IG, Huguet RB, Sousa LP, Abreu MN, Rocha NP, Bauer ME, Carvalho LA, Teixeira AL. Circulating levels of GDNF in bipolar disorder. Neuroscience Letters 2011;502(2) 103-106.

[74] Takebayashi M, Hisaoka K, Nishida A, Tsuchioka M, Miyoshi I, Kozuru T, Hikasa S, Okamoto Y, Shinno H, Morinobu S, Yamawaki S. Decreased levels of whole blood glial cell line-derived neurotrophic factor (GDNF) in remitted patients with mood disorders. The International Journal of Neuropsychopharmacology / Official Scientific Journal of the Collegium Internationale Neuropsychopharmacologicum (CINP) 2006;9(5) 607-612.

[75] McCusker RH, McCrea K, Zunich S, Dantzer R, Broussard SR, Johnson RW, Kelley KW. Insulin-like growth factor-I enhances the biological activity of brain-derived neurotrophic factor on cerebrocortical neurons. Journal of Neuroimmunology 2006;179(1-2) 186-90.

[76] Torres-Aleman I. Toward a comprehensive neurobiology of IGF-I. Developmental Neurobiology 2010;70(5) 384-396.

[77] Bibollet-Bahena O, Cui QL, Almazan G. The insulin-like growth factor-1 axis and its potential as a therapeutic target in central nervous system (CNS) disorders. Central Nervous System Agents in Medicinal Chemistry 2009;9(2) 95-109.

[78] Annunziata M, Granata R, Ghigo E. The IGF system. Acta Diabetol 2011;48(1) 1-9.

[79] Szczęsny E, Slusarczyk J, Głombik K, Budziszewska B, Kubera M, Lasoń W, Basta-Kaim A. Possible contribution of IGF-1 to depressive disorder. Pharmacological Reports: PR 2013;65(6) 1622-1631.

[80] Russo VC, Gluckman PD, Feldman EL, Werther GA. The insulin-like growth factor system and its pleiotropic functions in brain. Endocrine Reviews 2005;6 916-943.

[81] Aberg MA, Aberg ND, Hedbäcker H, Oscarsson J, Eriksson PS. Peripheral infusion of IGF-I selectively induces neurogenesis in the adult rat hippocampus. The Journal of neuroscience: the Official Journal of the Society for Neuroscience 2000;20(8) 2896-2903.

[82] Liu JP, Baker J, Perkins AS, Robertson EJ, Efstratiadis A. Mice carrying null mutations of the genes encoding insulin-like growth factor I (Igf-1) and type 1 IGF receptor (Igf1r). Cell 1993;75(1) 59-72.

[83] O'Kusky JR, Ye P, D'Ercole AJ. Insulin-like growth factor-I promotes neurogenesis and synaptogenesis in the hippocampal dentate gyrus during postnatal development. The Journal of Neuroscience: the Official Journal of the Society for Neuroscience 2000;20(22) 8435-8442.

[84] Cheng CM, Cohen M, Tseng V, Bondy CA. Endogenous IGF1 enhances cell survival in the postnatal dentate gyrus. Journal of Neuroscience Research 2001;64(4) 341-347.

[85] Mitschelen M, Yan H, Farley JA, Warrington JP, Han S, Hereñú CB, Csiszar A, Ungvari Z, Bailey-Downs LC, Bass CE, Sonntag WE.Long-term deficiency of circulating and hippocampal insulin-like growth factor I induces depressive behavior in adult mice: a potential model of geriatric depression. Neuroscience 2011;185 50-60.

[86] Lee KY, Miki T, Yokoyama T, Ueki M, Warita K, Suzuki S, Ohta K, Wang ZY, Jamal M, Yakura T, Liu JQ, Hosomi N, Takeuchi Y. Neonatal repetitive maternal separation causes long-lasting alterations in various neurotrophic factor expression in the cerebral cortex of rats. Life Sciences 2012;90(15-16) 578-584.

[87] Erabi K, Morinobu S, Kawano K, Tsuji S, Yamawaki S. Neonatal isolation changes the expression of IGF-IR and IGFBP-2 in the hippocampus in response to adulthood restraint stress. The International Journal of Neuropsychopharmacology / Official Scientific Journal of the Collegium Internationale Neuropsychopharmacologicum (CINP) 2007;10(3) 369-381.

[88] Hoshaw BA, Malberg JE, Lucki I.Central administration of IGF-I and BDNF leads to long-lasting antidepressantlike activity. Brain Research 2005;1037(1-2) 204-208.

[89] Hoshaw BA, Hill TI, Crowley JJ, Malberg JE, Khawaja X, Rosenzweig-Lipson S, Schechter LE, Lucki I. Antidepressant-like behavioral effects of IGF-1 produced by enhanced serotonin transmission. European Journal of Pharmacology 2008;594(1-3) 109-116.

[90] Park SE, Dantzer R, Kelley KW, McCusker RH.Central administration of insulin-like growth factor-I decreases depressive-like behavior and brain cytokine expression in mice. Journal of Neuroinflammation 2011;8 1-14.

[91] Grunbaum-Novak N, Taler M, Gil-Ad I, Weizman A, Cohen H, Weizman R. Relationship between antidepressants and IGF-1 system in the brain: possible role in cog-

nition. European neuropsychopharmacology: the Journal of the European College of Neuropsychopharmacology 2008;18(6) 431-8.

[92] Khawaja X, Xu J, Liang JJ, Barrett JE. Proteomic analysis of protein changes developing in rat hippocampus after chronic antidepressant treatment: Implications for depressive disorders and future therapies. Journal of neuroscience research 2004;75(4) 451-460.

[93] Weber-Haman B, Blum WF, Kratzsch J, Gilles M, Heuser I,Deuschle M.Insulin-like growth factor-I (IGF-I) serum concentrations in depressed patients: relationship to saliva cortisol and changes during antidepressant treatment. Pharmacopsychiatry 2009;42(1) 23-28.

[94] Franz B, Buysse DJ, Cherry CR, Gray NS, Grochocinski VJ, Frank E, Kupfer DJ. Insulin-like growth factor 1 and growth hormone binding protein in depression: a preliminary communication. Journal of Psychiatric Research 1999;33(2) 121-127.

[95] Sievers C, Auer MK, Klotsche J, Athanasoulia AP, Schneider HJ, Nauck M, Völzke H, John U, Schulz A, Freyberger HJ, Friedrich N, Biffar R, Stalla GK, Wallaschofski H, Grabe HJ. IGF-I levels and depressive disorders: results from the Study of Health in Pomerania (SHIP). European Neuropsychopharmacology: the Journal of the European College of Neuropsychopharmacology 2014; 24(6) 890-896.

[96] Alder J, Thakker-Varia S, Bangasser DA, Kuroiwa M, Plummer MR, Shors TJ, Black IB. Brain-derived neurotrophic factor-induced gene expression reveals novel actions of VGF in hippocampal synaptic plasticity. The Journal of Neuroscience: the Official Journal of the Society for Neuroscience 2003;23(34) 10800-10808

[97] Bonni A, Ginty DD, Dudek H, Greenberg ME. Serine 133-phosphorylated CREB induces transcription via a cooperative mechanism that may confer specificity to neurotrophin signals. Molecular and Cellular Neurosciences 1995;6(2) 168-183.

[98] Hahm S, Mizuno TM, Wu TJ, Wisor JP, Priest CA, Kozak CA, Boozer CN, Peng B, McEvoy RC, Good P, Kelley KA, Takahashi JS, Pintar JE, Roberts JL, Mobbs CV, Salton SR. Targeted deletion of the Vgf gene indicates that the encoded secretory peptide precursor plays a novel role in the regulation of energy balance. Neuron 1999; 23(3) 537-548.

[99] Snyder SE, Salton SR. Expression of VGF mRNA in the adult rat central nervous system. The Journal of Comparative Neurology 1998;394(1) 91-105.

[100] Cattaneo A, Sesta A, Calabrese F, Nielsen G, Riva MA, Gennarelli M. The expression of VGF is reduced in leukocytes of depressed patients and it is restored by effective antidepressant treatment. Neuropsychopharmacology: Official Publication of the American College of Neuropsychopharmacology 2010;35(7) 1423-1428.

[101] Thakker-Varia S, Krol JJ, Nettleton J, Bilimoria PM, Bangasser DA, Shors TJ, Black IB, Alder J. The neuropeptide VGF produces antidepressant-like behavioral effects and

enhances proliferation in the hippocampus. The Journal of neuroscience: the Official Journal of the Society for Neuroscience 2007;27(45) 12156-12167.

[102] Hunsberger JG, Newton SS, Bennett AH, Duman CH, Russell DS, Salton SR, Duman RS. Antidepressant actions of the exercise-regulated gene VGF. Nature Medicine 2007;13(12) 1476-1482.

[103] Altar CA, Laeng P, Jurata LW, Brockman JA, Lemire A, Bullard J, Bukhman YV, Young TA, Charles V, Palfreyman MG. Electroconvulsive seizures regulate gene expression of distinct neurotrophic signaling pathways. The Journal of Neuroscience: the Official Journal of the Society for Neuroscience 2004;24(11) 2667-2677.

[104] Severini C, Ciotti MT, Biondini L, Quaresima S, Rinaldi AM, Levi A, Frank C, Possenti R. TLQP-21, a neuroendocrine VGF-derived peptide, prevents cerebellar granule cells death induced by serum and potassium deprivation. Journal of Neurochemistry 2008;104(2) 534-544.

[105] Malberg JE, Eisch AJ, Nestler EJ, Duman RS. Chronic antidepressant treatment increases neurogenesis in adult rat hippocampus. The Journal of neuroscience: the Official Journal of the Society for Neuroscience 2000;20(24) 9104-9110.

[106] Newton SS, Collier EF, Hunsberger J, Adams D, Terwilliger R, Selvanayagam E, Duman RS. Gene profile of electroconvulsive seizures: induction of neurotrophic and angiogenic factors. The Journal of neuroscience: the Official Journal of the Society for Neuroscience 2003;23(34) 10841-10851.

[107] Cattaneo A, Sesta A, Calabrese F, Nielsen G, Riva MA, Gennarelli M. The expression of VGF is reduced in leukocytes of depressed patients and it is restored by effective antidepressant treatment. Neuropsychopharmacology : official publication of the American College of Neuropsychopharmacology 2010; 35(7) 1423-1428.

[108] Cattaneo A, Gennarelli M, Uher R, Breen G, Farmer A, Aitchison KJ, Craig IW, Anacker C, Zunsztain PA, McGuffin P, Pariante CM. Candidate genes expression profile associated with antidepressants response in the GENDEP study: differentiating between baseline 'predictors' and longitudinal 'targets'. Neuropsychopharmacology : official publication of the American College of Neuropsychopharmacology 2013;38(3) 377-385.

[109] Thakker-Varia S, Jean YY, Parikh P, Sizer CF, Jernstedt Ayer J, Parikh A, Hyde TM, Buyske S, Alder J. The neuropeptide VGF is reduced in human bipolar postmortem brain and contributes to some of the behavioral and molecular effects of lithium. The Journal of Neuroscience: the Official Journal of the Society for Neuroscience 2010;30(28) 9368-9380.

[110] Huang JT, Leweke FM, Oxley D, Wang L, Harris N, Koethe D, Gerth CW, Nolden BM, Gross S, Schreiber D, Reed B, Bahn S. Disease biomarkers in cerebrospinal fluid of patients with first-onset psychosis. PLoS Medicine 2006;3(11) 2145-2158.

[111] Turner CA, Akil H, Watson SJ, Evans SJ. The Fibroblast Growth Factor System and Mood Disorders. Biological Psychiatry 2006;59(12) 1128-1135.

[112] Turner CA, Gula EL, Taylor LP, Watson SJ, Akil H. Antidepressant-Like Effects of Intracerebroventricular FGF2 in Rats. Brain Research 2008;1224 63-68.

[113] Yoshimura S, Teramoto T, Whalen MJ, Irizarry MC, Takagi Y, Qiu J, Harada J, Waeber C, Breakefield XO, Moskowitz MA. FGF-2 regulates neurogenesis and degeneration in the dentate gyrus after traumatic brain injury in mice. The Journal of Clinical Investigation 2003;112(8) 1202-1210.

[114] Evans SJ, Choudary PV, Neal CR, Li JZ, Vawter MP, Tomita H, Lopez JF, Thompson RC, Meng F,Stead JD, Walsh DM, Myers RM, Bunney WE, Watson SJ, Jones EG, Akil H. Dysregulation of the fibroblast growth factor system in major depression. Proceedings of the National Academy of Sciences of the United States of America 2004;101(43) 15506-15511.

[115] He S, Zhang T, Hong B, Peng D, Su H, Lin Z, Fang Y, Jiang K, Liu X, Li H. Decreased serum fibroblast growth factor-2 levels in pre-and post-treatment patients with major depressive disorder. Neuroscience Letters 2014;579 168-72.

[116] Cheng Y, Black IB, DiCicco-Bloom E. Hippocampal granule neuron production and population size are regulated by levels of bFGF. The European Journal of Neuroscience 2002;15(1) 3-12.

[117] Ohkubo Y, Uchida AO, Shin D, Partanen J, Vaccarino FM. Fibroblast growth factor receptor 1 is required for the proliferation of hippocampal progenitor cells and for hippocampal growth in mouse. The Journal of Neuroscience: the Official Journal of the Society for Neuroscience 2004;24(27) 6057-6069.

[118] Raballo R, Rhee J, Lyn-Cook R, Leckman JF, Schwartz ML, Vaccarino FM. Basic fibroblast growth factor (Fgf2) is necessary for cell proliferation and neurogenesis in the developing cerebral cortex. The Journal of Neuroscience: the Official Journal of the Society for Neuroscience 2000;20(13) 5012-5023.

[119] Yoshimura S, Takagi Y, Harada J, Teramoto T, Thomas SS, Waeber C, Bakowska JC, Breakefield XO, Moskowitz MA. FGF-2 regulation of neurogenesis in adult hippocampus after brain injury. Proceedings of the National Academy of Sciences of the United States of America 2001;98(10) 5874-5879.

[120] Elsayed M, Banasr M, Duric V, Fournier NM, Licznerski P, Duman RS. Antidepressant effects of fibroblast growth factor-2 in behavioral and cellular models of depression. Biological Psychiatry 2012;72(4) 258-265.

[121] Bernard R, Kerman IA, Thompson RC, Jones EG, Bunney WE, Barchas JD, Schatzberg AF, Myers RM, Akil H, Watson SJ. Altered expression of glutamate signaling, growth factor, and glia genes in the locus coeruleus of patients with major depression. Molecular Psychiatry 2011;16 634-646.

[122] Kato M, Okugawa G, Wakeno M, Takekita Y, Nonen S, Tetsuo S, Nishida K, Azuma J, Kinoshita T, Serretti A. Effect of basic fibroblast growth factor (FGF2) gene polymorphisms on SSRIs treatment response and side effects. European neuropsychopharmacology: the Journal of the European College of Neuropsychopharmacology 2009;19 718-725.

[123] Bothwell M. Functional interactions of neurotrophins and neurotrophin receptors. Annual Review of Neuroscience 1995;18 223-253.

[124] Shimazu K, Zhao M, Sakata K, Akbarian, S, Bates B, Jaenisch R, Lu B. NT-3 facilitates hippocampal plasticity and learning and memory by regulating neurogenesis. Learning & Memory (Cold Spring Harbor, N.Y.) 2006;13(3) 307-315.

[125] Sheline YI, Wang PW, Gado MH, Csernansky JG, Vannier MW. Hippocampal atrophy in recurrent major depression. Proceedings of the National Academy of Sciences of the United States of America 1996;93(9) 3908-3913.

[126] Arenas E, Persson H. Neurotrophin-3 prevents the death of adult central noradrenergic neurons in vivo. Nature 1994;367 368-371.

[127] Leonard BE. The role of noradrenaline in depression: a review. Journal of Psychopharmacology (Oxford, England) 1997;11(4 suppl) S39-S47.

[128] Schutte A, Yan Q, Mestres P, Giehl KM. The endogenous survival promotion of axotomized rat corticospinal neurons by brain-derived neurotrophic factor is mediated via paracrine, rather than autocrine mechanisms. Neuroscience Letters 2000;290(3) 185-188.

[129] Paul J, Gottmann K, Lessmann V. NT-3 regulates BDNF-induced modulation of synaptic transmission in cultured hippocampal neurons. Neuroreport 2001;12(12) 2635-2639.

[130] Martin-Iverson MT, Todd KG, Altar CA. Brain-derived neurotrophic factor and neurotrophin-3 activate striatal dopamine and serotonin metabolism and related behaviors: Interactions with amphetamine. Journal of Neuroscience 1994;14(3 Pt 1) 1262-1270.

[131] Sieber-Blum M, Ren Z. Norepinephrine transporter expression and function in noradrenergic cell differentiation. Molecular and Cellular Biochemistry 2000;212(1–2) 61-70.

[132] Stanley SC, Brooks SD, Butcher JT, d'Audiffret AC, Frisbee SJ, Frisbee JC. PROTECTIVE EFFECT OF SEX ON CHRONIC STRESS-AND DEPRESSIVE BEHAVIOR-INDUCED VASCULAR DYSFUNCTION IN BALB/cJ MICE. Journal of Applied Physiology (Bethesda, Md. : 1985) 2014;pii: jap.00537.2014.

[133] Smith MA, Makino S, Kvetnansky R, Post RM. Stress and Glucocorticoids Affect the Expression of Brain-Derived Neurotrophic Factor and Neurotrophin-3 mRNAs in the

Hippocampus. The Journal of neuroscience: the Official Journal of the Society for Neuroscience 1995;15(3 Pt 1) 1766-1777.

[134] Kushikata T, Kubota T, Fang J, Krueger JM. Neurotrophins 3 and 4 enhance non-rapid eye movement sleep in rabbits. Neuroscience Letters 2003;346(3) 161-164.

[135] Walz JC, Frey BN, Andreazza AC, Ceresér KM, Cacilhas AA, Valvassori SS, Quevedo J, Kapczinski F. Effects of lithium and valproate on serum and hippocampal neurotrophin-3 levels in an animal model of mania. Journal of Psychiatric Research 2008;42 (5) 416-421.

[136] Ghosh A, Greenberg ME. Distinct roles for bFGF and NT-3 in the regulation of cortical neurogenesis. Neuron 1995;15 89-103.

[137] Walz JC, Andreazza AC, Frey BN, Cacilhas AA, Ceresér KM, Cunha AB, Weyne F, Stertz L, Santin A, Gonçalves CA, Kapczinski F. Serum neurotrophin-3 is increased during manic and depressive episodes in bipolar disorder. Neuroscience Letters 2007;415(1):87-89.

[138] Fernandes BS, Gama CS, Walz JC, Ceresér KM, Fries GR, Colpo G, Aguiar B, Pfaffenseller B, Kauer-Sant'anna M, Kapczinski F. Increased neurotrophin-3 in drug-free subjects with bipolar disorder during manic and depressive episodes. Journal of Psychiatric Research 2010;44(9):561-565.

[139] Hock C, Heese K, Müller-Spahn F, Huber P, Riesen W, Nitsch RM, Otten U. Increased cerebrospinal fluid levels of neurotrophin 3 (NT-3) in elderly patients with major depression. Molecular Psychiatry 2000;5(5) 510-513.

[140] Nockher WA, Renz H. Neurotrophins and asthma: novel insight into neuroimmune interaction. The Journal of Allergy and Clinical Immunology 2006;117(1): 67-71.

[141] Covaceuszach S, Capsoni S, Ugolini G, Spirito F, Vignone D, Cattaneo A. Development of a non invasive NGF-based therapy for Alzheimer's disease. Current Alzheimer Research 2009;6(2) 158-170.

[142] Cohen S. Purification of a nerve-growth promoting protein from the mouse salivary gland and its neuro-cytotoxic antiserum. Proceedings of the National Academy of Sciences of the United States of America 1960;46 302-311.

[143] Schulte-Herbrüggen O, Chourbaji S, Müller H, Danker-Hopfe H, Brandwein C, Gass P, Hellweg R. Differential regulation of nerve growth factor and brain-derived neurotrophic factor in a mouse model of learned helplessness. Experimental Neurology 2006;202(2) 404-409.

[144] Schulte-Herbrüggen O, Fuchs E, Abumaria N, Ziegler A, Danker-Hopfe H, Hiemke C, Hellweg R. Effects of escitalopram on the regulation of brain-derived neurotrophic factor and nerve growth factor protein levels in a rat model of chronic stress. Journal of Neuroscience Rresearch 2009;87(11) 2551-2560.

[145] Della FP, Abelaira HM, Réus GZ, Santos MA, Tomaz DB, Antunes AR, Scaini G, Morais MO, Streck EL, Quevedo J. Treatment with tianeptine induces antidepressive-like effects and alters the neurotrophin levels, mitochondrial respiratory chain and cycle Krebs enzymes in the brain of maternally deprived adult rats. Metabolic Brain Disease 2013;28(1) 93-105.

[146] de Azevedo Cardoso T, Mondin TC, Wiener CD, Marques MB, Fucolo Bde Á, Pinheiro RT, de Souza LD, da Silva RA, Jansen K, Oses JP. Neurotrophic factors, clinical features and gender differences in depression. Neurochemical Research 2014;39(8) 1571-1578.

[147] Liu X, Zhang T, He S, Hong B, Peng D, Su H, Li F, Tang Y, Lin Z, Fang Y, Jiang K. Nerve growth factor variations in patients with mood disorders: no changes in eight weeks of clinical treatment. Neuropsychiatric Disease and Ttreatment 2014;10 835-840.

[148] Cirulli F, Alleva E. The NGF saga: from animal models of psychosocial stress to stress-related psychopathology. Frontiers in Neuroendocrinology 2009(a);30(3) 379-395.

[149] Cirulli F, Francia N, Branchi I, Antonucci MT, Aloe L, Suomi SJ, Alleva E. Changes in plasma levels of BDNF and NGF reveal a gender-selective vulnerability to early adversity in rhesus macaques. Psychoneuroendocrinology 2009(b);34(2) 172-180.

[150] Alleva E, Petruzzi S, Cirulli F, Aloe L. NGF regulatory role in stress and coping of rodents and humans. Pharmacology, Biochemistry, and Behavior 1996;54(1) 65-72.

[151] Xiong P, Zeng Y, Wan J, Xiaohan DH, Tan D, Lu J, Xu F, Li HY, Zhu Z, Ma M. The role of NGF and IL-2 serum level in assisting the diagnosis in first episode schizophrenia. Psychiatry Research 2011;189(1) 72-76.

[152] Pallavi P, Sagar R, Mehta M, Sharma S, Subramanium A, Shamshi F, Sengupta U, Qadri R, Pandey RM, Mukhopadhyay AK. Serum neurotrophic factors in adolescent depression: gender difference and correlation with clinical severity. Journal Affective Disorders 2013;150 415–423.

[153] Martino M, Rocchi G, Escelsior A, Contini P, Colicchio S, de Berardis D, Amore M, Fornaro P, Fornaro M. NGF serum levels variations in major depressed patients receiving duloxetine. Psychoneuroendocrinology 2013;38(9) 1824-1828.

[154] Diniz BS, Teixeira AL, Machado-Vieira R, Talib LL, Gattaz WF, Forlenza OV. Reduced serum nerve growth factor in patients with late-life depression. The American Journal of Geriatric Psychiatry: Official Journal of the American Association for Geriatric Psychiatry 2013;21(5) 493-496.

[155] Hadjiconstantinou M, McGuire L, Duchemin AM, Laskowski B, Kiecolt-Glaser J, Glaser R. Changes in plasma nerve growth factor levels in older adults associated with chronic stress. Journal of Neuroimmunology 2001;116 102-106.

[156] Fontenelle LF, Barbosa IG, Luna JV, Rocha NP, Miranda AS, Teixeira AL. Neurotrophic fators in obsessive-compulsive disorder. Psychiatry Research 2012;199 195-200.

[157] Barbosa IG, Huguet RB, Neves FS, Reis HJ, Bauer ME, Janka Z, Palotás A, Teixeira AL. Impaired nerve growth factor homeostasis in patients with bipolar disorder. The world journal of biological psychiatry: the Official Journal of the World Federation of Societies of Biological Psychiatry 2011;12 228-232.

[158] Conti G, Gale K, Kondratyev A. Immunohistochemical evaluation of the protein expression of nerve growth factor and its TrkA receptor in rat limbic regions following electroshock seizures. Neuroscience Research 2009;65(2) 201-209.

[159] Bilgen AE, Bozkurt Zincir S, Zincir S, Ozdemir B, Ak M, Aydemir E, Sener I. Effects of electroconvulsive therapy on serum levels of brain-derived neurotrophic factor and nerve growth factor in treatment resistant major depression. Brain Research Bulletin 2014;104 82-87.

[160] Grønli O, Stensland GØ, Wynn R, Olstad R. Neurotrophic factors in serum following ECT: a pilot study. The World Journal of Biological Psychiatry: the Official Journal of the World Federation of Societies of Biological Psychiatry 2009;10(4) 295-301.

Chapter 3

The Molecular and Quantum Approach to Psychopathology and Consciousness — From Theory to Experimental Practice

Massimo Cocchi, Lucio Tonello and Fabio Gabrielli

Additional information is available at the end of the chapter

http://dx.doi.org/10.5772/59392

> *With the idol of certainty [...] there falls one of the defences of obscurantism [...] for the worship of this idol hampers not only the boldness of the questions, but also the rigor and the integrity of our tests. The wrong view of science betrays itself in the craving to be right; for it is not its possession of knowledge [...] that makes the man of science, but its persistent and recklessly critical quest for truth.*(K.R.Popper, *The Logic of Scientific Discovery*, engl. tr. Hutchinson, London 1959)

1. Introduction

This decade has clocked the review of the new DSM, the fifth in the series, the instrument considered the "bible" of psychiatry worldwide.

The document, which is accomplished today, appears firmly rooted in traditional conservative psychiatry ignoring the progress made by the biological research field. Clearly, the dichotomy between conservative and progressive psychiatry is not over, despite the efforts of the scientific research in the field of psychiatry, of brain, of neurotransmitters and of quantum computation of the brain and consciousness, i.e., the disciplines that belong to neuroscience. It seems correct, from the point of view of ethics, remember how it is difficult to think of the research in psychiatry as completely independent of influential external factors (kuhnian paradigms).

Recently some major events have allowed a movement of thought not only innovative, but of profound and insistent criticism, mainly at a high intellectual and scientific level, on the ideological implications of psychiatric diagnosis and of the increasing complexity of the nuances that classify the psychiatric disorder, rather than looking at a window that allows, through biological markers, a reliable diagnosis and appropriate care in the first diagnostic instance by limiting the diagnostic error unaware that psychiatric diagnosis has dragged on for years about the recognition of bipolar disorder from major depressive disorder [1] where there is a diagnostic misinterpretation ranging from 40% [1] to 70% (Tenth World Day for the Prevention of Suicide, Rome, 2012).

The fifth edition has been criticized by a number of authorities, even before it was formally published. The main thrust of criticism has been that changes in the DSM have not kept pace with advances in scientific understanding of psychiatric dysfunction. Another criticism is that the development of DSM-5 was unduly influenced by input from the psychiatric drug industry. A number of scientists have objected that the DSM forces clinicians to make distinctions that are not supported by solid evidence, distinctions that have major treatment implications, including drug prescriptions and the availability of health insurance coverage.

2. Retrospective research on humans and animals

In the first experimental phase two mathematical tools were identified, one complex (the Self-Organizing Map-SOM) and a simple one (the Index B2) which in time will prove valuable not only to define the condition of the Major Depression and Bipolar Disorder, but will provide the possibility of reasonable inferences about the biological significance of the two molecular mood disorders.

The SOM is an artificial neural network that has the ability to put together similar objects and distant different objects using the characteristics of the objects considered.

The index B2 (so named by the authors of the research, (namely, Cocchi and Tonello) is derived from a mathematical operation that relates the characteristics of molecular weight and melting point of the fatty acids isolated and recognized, by the SOM, to have the power of recognizing the two disorders.

Proceeding by grades we will say that the two groups of subjects investigated in the first phase of the research (apparently normal and depressed) were determined by the fatty acids of platelets, having chosen this cell type for the morphological and functional particularities that distinguish these cells, i.e., the presence of receptors for neurotransmitters, particularly serotonin, and because they are also the seat of the molecular events that regulate the hemocoagulation process.

The results obtained experimentally, interpreted by the non-linear mathematical function, the SOM, and the index B2 showed the ability to distinguish "psychiatric" patients from "normal" ones. The problem was that the psychiatric diagnosis we had received was generally expressed as Major Depression, while the arrangement of subjects within the framework resulting from the SOM and the index B2 induced us to think that some aspect was unclear about the diagnosis that we had received.

Placement in the SOM and the evaluation indexes B2 (negative and positive) led us to look for an opportunity to relaunch the experiment that, as we thought, might be able to recognize the apparently normal individuals from major depressive and bipolar.

The opportunity came with a grant from the Marche Region, and in two years of intense work, including the contributions of psychiatrists, biochemists, molecular biologists, mathematicians and quantum physicists. A research that has used a combination of biology and nonlinear mathematics was carried out, in order to identify, within the psychiatric chapter of mood disorders, whether it was possible to identify in the platelets, and in particular in their fatty acids, molecular features that could allow a clear and precise classification of subjects with Major Depression (MD) and with Bipolar Disorder (BD). The results were obtained using an Artificial Neural Network, in particular a network called Self-Organizing Map (SOM) [2, 3]. The SOM is an unsupervised competitive-learning network algorithm, which was created by Teuvo Kohonen in 1981–82 [4-6].

With the above combination, using platelet's Palmitic Acid, Linoleic Acid, and Arachidonic Acid together with SOM and a mathematic index (B2), it was possible to obtain the effect of discriminating between MD and BD, for the first time in years. The B2 index was obtained from the summation of the percentages of each fatty acid multiplied by its melting point and divided by its molecular weight, obtaining an indirect expression of membrane viscosity, which induces us to identify it with the neuron membrane viscosity [7]. The B2 index is found to be negative in MD and positive in BD, that is the membranes in MD are, by far, less viscous than in normals, in BD, in psychotics, showing a unique and specific molecular characteristic for subjects with MD [8]. On these bases it was possible to explain the quantitative biomolecular approach to major depression and hypothesize that in mood disorders a biomolecular pathway exists, moving from cell membrane viscosity through Gsα protein and Tubulin [9]. We got the result so desired and hoped to confirm that the guess was correct, the depressed subjects were distinguished from bipolar, beyond all psychiatric, classificatory and interpretative dialectics. Figures 1, 2, 3, 4, 5 summarize the main steps of the research. In front of the results, we began to reflect on the distribution and on the logic of the numbers that had been so intimately associated with psychiatric conditions, as they represented a fact which did not take into account therapies and nothing that from the outside could be related to subject. All this led us to think that there was something already written in platelets that can simulate the condition of the neuron, at least, as regards the levels of serotonin. On this observation we wrote some articles that related to the uniqueness of the molecular Major Depression, the role of membrane viscosity and molecular reflections on the state of consciousness and the mechanical strength of the membrane.

During the last experiment [10] the psychiatrists have provided us with eight cases of "Suicidal Ideation".

When we have classified them over the SOM, the Figure 6 was obtained.

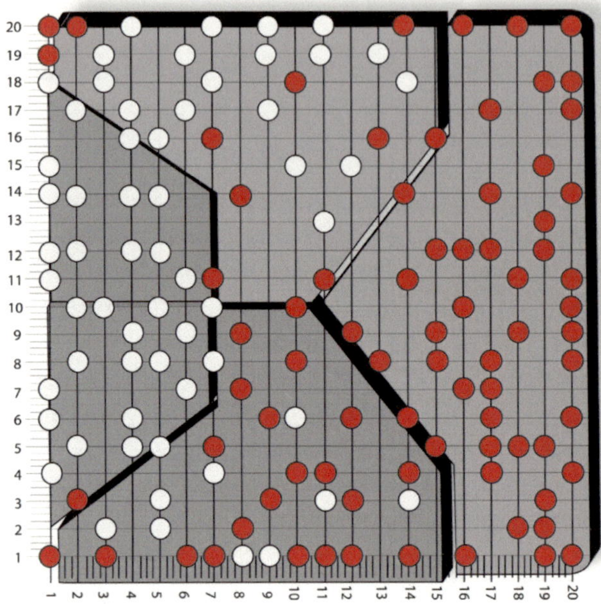

Figure 1. Distribution of all cases [apparently Normal (white) and Pathologic (red) over the SOM obtained by Platelets' Palmitic Acid (PA), Linoleic Acid (LA) and Arachidonic Acid (AA).

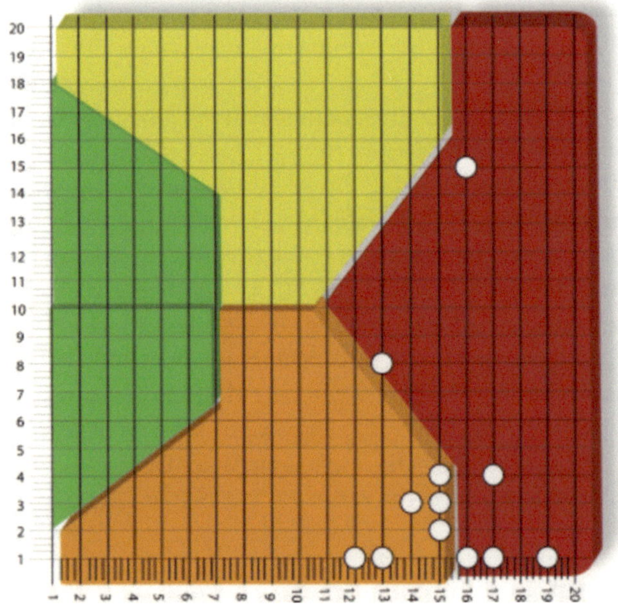

Figure 2. shows the pathologic subjects Major Depression and Bipolar Disorder) all together.

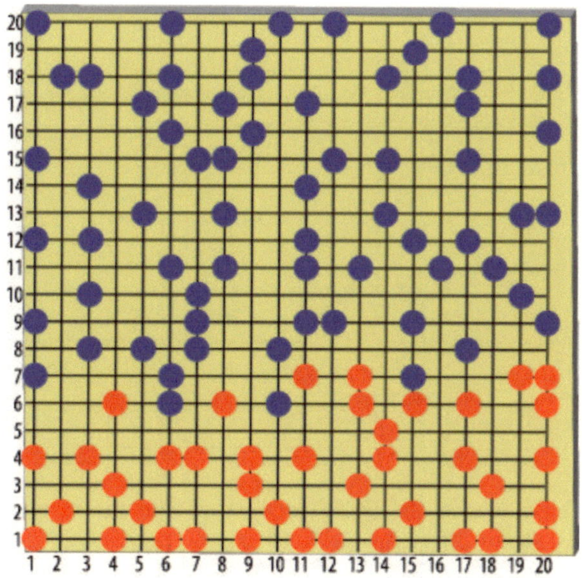

Figure 3. (a new SOM has been realized) shows, clearly, that it was possible to distinguish the subjects with Major Depression (red) from those one with Bipolar Disorder (blue).

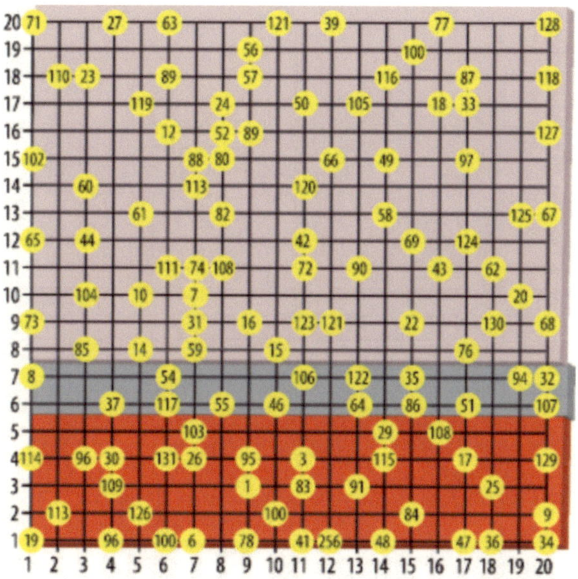

Figure 4. shows the same picture of figure 3 pointing out an intermediate area which collects both cases (Major Depression and Bipolar Disorder). For each subject we have calculated an Index called B2, We have obtained the B2 Index by the sum of the percentages of each fatty acids (AA, LA and PA), multiplied for the melting point and divided for the molecular weight. B2 is negative for subjects with Major Depression, positive for subjects with Bipolar Disorder. In this way it is possible to recognize also the cases that are within a very close range as showed in Figure 5.

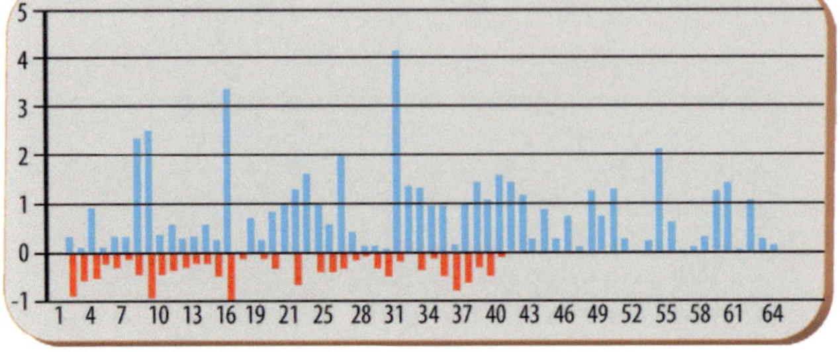

Figure 5. In Figure 5 is represented the classification of the subjects with Major Depression (B2 negative-red) and Bipolar Disorder (B2 positive-blue). As can be seen, the cases, also within a very close range, are clearly distinguishable. The combination of the SOM and of the B2 index is able to perform the right diagnosis [10].

The cases were collected where the SOM recognizes the minimum of Linoleic Acid. In particular seven cases were Bipolar and one with Major Depression, confirming that both can have suicidal ideation and can attempt suicide [10]. The subject, in position 15:4, was uncertain at the psychiatric evaluation; in effect his position is a little bit out from the critical area of the minimum of Linoleic Acid. In the same way, other areas has been found within the SOM (fig. 7): Obsessive Compulsive Disorder (OCD) area, Major Depression area, Bipolar area (the largest), Psychotic area etc.

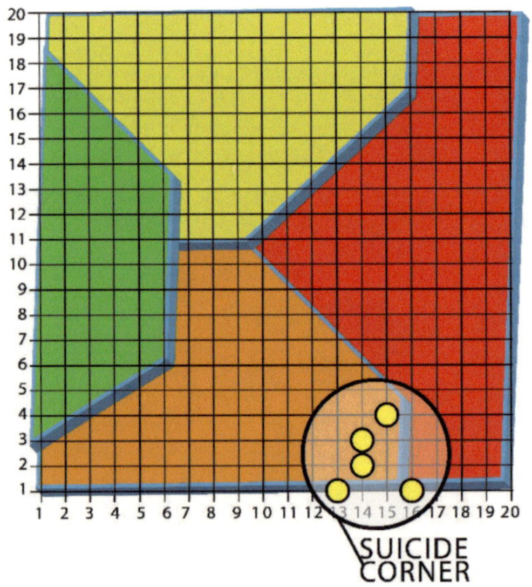

Figure 6. Distribution of the "suicidal" cases over the SOM.

All the experimental findings in humans and animals are resumed in Figures 7 and 8.

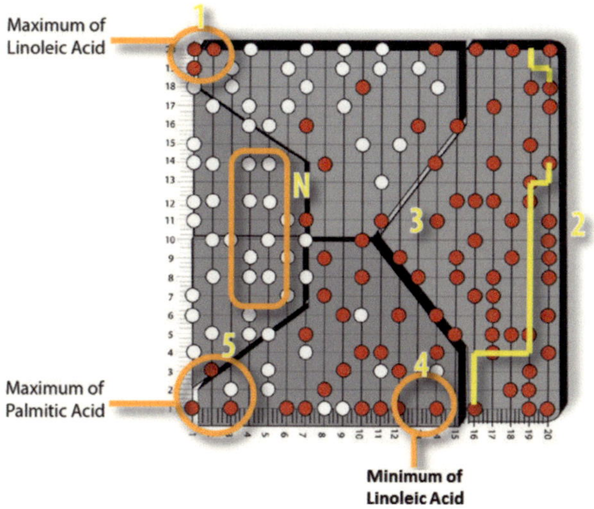

Figure 7. Distribution over the SOM of human subjects and animals. According to the psychiatric diagnosis (when definitive) we can recognize: 1= OCD area, 2= Major Depression area, 3= Bipolar area (the largest), 4= Suicide area, 5= Psychotic area, N= apparently normal area. N area collects about the 50% of the sample of subjects considered apparently normal.

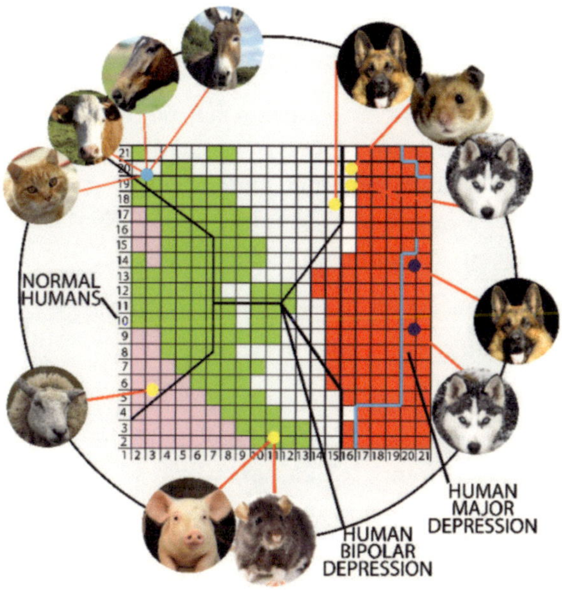

Figure 8. Distribution over the SOM of human subjects and animals. According to the psychiatric diagnosis (when definitive) we can recognize: 1= OCD area, 2= Major Depression area, 3= Bipolar area (the largest), 4= Suicide area, 5= Psychotic area, N= apparently normal area. N area collects about the 50% of the sample of subjects considered apparently normal.

Several different animals have been mapped on the SOM, as well. The molecular similarities [11], observed between animal and man (Figure 8), concern the conditions of Major Depression, Bipolar Disorder, Obsessive Compulsive Disorder (OCD).

To confirm the molecular correspondence between man and animal, observe how, e.g. Cat, Bovine, Horse and Donkey, correspond to the area of maximum Linoleic Acid and of Obsessive Compulsive Disorder.

This area is recognized as the point of maximum concentration of Linoleic not only for diagnostics correspondence, but also because it contains the cat, who, as feline, is known to possess desaturase, but with low activity [12], therefore not to be able to transform Linoleic Acid into Arachidonic Acid, resulting in savings of Linoleic, and long living animals [13]. Further, in the same animals, symptoms of OCD can occur [14, 15]. **See Appendix (Linoleic acid secrets).**

3. On the non-manipulability of the SOM built for the classification of the psychiatric subjects

Let's suppose we want to build a fake SOM, that is, a SOM driven by us according to a desired result. We should be very lucky, in fact we should guess:

1200 particular numbers (starting weights). By the way, really, it is impossible to know how they could be chosen in order to obtain a particular result.

Above all, we should find a particular order of data that, because of an unknown reason (really unknown), lead to a very particular result. In our case, we have a data base of 144 Subjects (84 depressive and 60 normal). This means that there are 144!=5.5503*10249 combinations. A training process takes about 4 minutes. So, we need about 4.224*10242 centuries to check all possible results thus taking a particular one. A bit difficult.

So, if we want to build a fake SOM, well, it's almost impossible (at least in a reasonable time even using the fastest computer on Earth).

The SOM (Figure 1) shows that major depressive subjects belong to an area which is completely disconnected from that of healthy and bipolar. Looking at the location of the data over the SOM, we find also a region (extreme left corner) which we attribute to psychotic subjects according to the clinical diagnosis. We translated these facts in terms of symmetry breaking [16], confirming that MD is a disease completely trapped apart from healthy, bipolar, psychotic subjects and patients with obsessive compulsive disorder (OCD) (Figure 9).

"What is opposition is reconciled, and by different things the more beautiful harmony is created, and everything is generated by the contrast."

Heraclitus

http://dx.doi.org/10.5772/59392

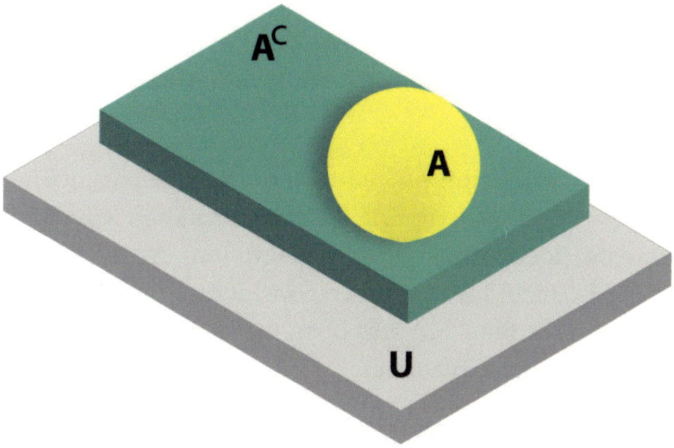

Figure 9. Bipartition of U. U is the Universal set representing humankind. A is the cell whose elements are characterized by a positive value of B2. A^C, which is the complement of A in U, is the cell whose elements are characterized by a negative value of B2. B2 is the index who put in relation the three fatty acids, isolated by the SOM, with the molecular weight and the melting point and that has recognized the Bipolar subjects (B2 positive) from the Major Depressive subjects (B2 negative).

4. Facts and perspectives on quantum neuron molecular research

The Quantum Paradigm Psychopathology Group (QPP) conference in Palermo (26-27 of April, 2013) has marked a definite turning point in the foundational perspective of many of the group's participants regarding the study of psychopathology, particularly mood disorders. One reason for this turning point stems from a realization that two of the most common forms of psychopathology, major depression and bipolar disorder, may be recognizable through bimolecular markers (**see Appendix: from biology to the anthropology of treatment**). Long years of theoretical study – out of the conviction that one should not be, using Feyerabend's words, "thought officer and concept manager", but rather, as Lakatos claims, good creator of theoretical frameworks able of acting "faster than the records of facts that must be collected in them"-by independent investigators have finally culminated in a convergence of their insights though quantum paradigms that now promise to illuminate, through the empirically tangible route of such new bio molecular markers, pathological phenomena of the conscious brain, thus potentially both factually confirming and further harmonizing the diverse prior contributions of these conceptually innovative psychiatrists, biochemists, molecular biologists, philosophers and theologians.

The idea, as stressed during the Conference in Palermo, was to take into consideration consciousness processes, together with their normal and pathological dynamics, without fixating on isolated elements and levels that can be observed by a privileged and detached observer, but rather by thinking and acting effectively through connections, relations, and networks. And all of this, always in the belief that science is the narration of a world that expresses emerging states, and is therefore never reducible to a simple sum of basics ingredients, where spontaneous symmetry breakings ensure multiplicity, creativity, vitality, in compliance with the concept of natural self-organization and systems evolution, towards growing complex and unpredictable states.

The socio-economic significance of this procedure is undeniable: science takes shape into social and economic structures which, by accepting the transformation from foucaultian monitoring and control instruments (ideological reductionism) into open, fluid, emerging systems (complexity or open logics), could really, when considering mental diseases, understand the often blurred classifications of the DSM and open up to the important connections between consciousness and quantum brain dynamics.

Hence the rejection of any form of ontological reductionism, which is self-referring, linked to metaphysical-ideological cognitive dynamics, and tied to a pervasive will to power which sees research freedom as a worrying system breakdown.

Against this epistemological backdrop, among the foundational innovators we can mention those who have left particularly fertile footprints in terms of basic quantum theories linking brain, behaviour, and consciousness.

Quantum Mind has been an ongoing field of study since the final decades of the last century. Pioneers like the physicists Hiroomi Umezawa, Kunio Yasue, and Giuseppe Vitiello, mathematicians like Roger Penrose, and biomedical investigators like Stuart Hameroff, Gordon Globus, and Gustav Bernroider have plumbed the depths of subatomic structure and its macroscopic amplifications in search of substrates for quantum computation and other capabilities that may match attributes of the normal human psyche better than models advocated by conventional cognitive neuroscience.

In the domain of psychopathology, Gordon Globus has gone on to propound a highly original concept of schizophrenia linked to the "tuning" of quantum vibrations suffusing the brain. Nancy Woolf, along with co-authors including Jack Tuszynski, has offered credible links between psychopathology and quantum-computational dysfunction within the skeletal proteins giving shape to brain cells. Paavo Pylkkanen has related the physical substrates of mental illness to quantum "pilot waves." Donald Mender has proposed ways of comprehending the neurophysiology of disordered thinking and emotion in terms of quantum "phase transitional" analogies to the freezing and melting of ordinary matter; he has also contributed to a reframing of psychiatric disease nosology in light of the anthropic principle. Ursula Werneke has complimented this anthropic reconsideration through her examination of psychotically "impaired" reality-testing in the context of Hugh Everett's many-worlds ontology. Massimo Cocchi, Lucio Tonello, Fabio Gabrielli, have forged links between serotonin and quantum phenomena via membrane biophysics in depression and psychosis.

The above honor roll of seminal QPP theoreticians is surely not exhaustive, but these brief remarks are not intended as a complete historical review. Rather, the purpose is an opening into the possibility of turning today's theoretical potentialities into experimental confirmed reality. It should be recalled and emphasized as a guiding principle that the cohesion of a convivial multidisciplinary group operating without the winnowing constraints of competing, mutually exclusive ideas may not remain true to the epistemic rigors of science. QPP can minimize this sort of hazard by maximizing, in the spirit of Karl Popper, exposure of its most cherished conjectures to a fair risk of experimental refutation.

A number of participants in the Palermo conference have signed a document, aptly called "The Declaration of Palermo," whose conclusion asserts:

> *"Even the absence of highly complex synaptic connections among neurons does not preclude the presence of at least rudimentary phenomenal experience in organisms endowed with superposed micro tubular dimers, ordered water, membrane ion channels, and/or crucial lipid raft assemblies connected to selected second messenger systems. In addition, quantum-biophysical aspects of these and/or other yet undiscovered structures and related processes may prove to be potent factors in the deeper etiologies and improved treatments of psychiatric disorders."*
>
> The Declaration of Palermo was written by Donald Mender and Massimo Cocchi and edited by: Don Michele Aramini, Gustav Bernroider, Francesco Cappello, Fabio Gabrielli, Gordon Globus, Mansoor Malik, Efstratios Manousakis, Kary Mullis, Eliano Pessa, Massimo Pregnolato, Paavo Pylkkänen, Mark M. Rasenick, Lucio Tonello, Jack Tuszynski, Giuseppe Vitiello, Ursula Werneke, Paola Zizzi.

This strong theoretical statement invites an opening into possible experimental models that will test the reality of the group's hypotheses by identifying, starting from precise molecular reference points characterizing the two mood disorders mentioned above, a non-trivially quantum pathway of bio molecular changes conditioning brain processes through the most intimate aspects of neuronal, trans neuronal, and sub neuronal function. In particular, membrane viscosity and its role within the interactome may prove to figure centrally in quantum-chemical transduction of neural signals.

The Declaration of Palermo concerning the plausibility of a quantum basis for consciousness entails a lucid analysis of phenomenologies crucial to both human beings and other creatures. The main feature belonging intrinsically to both Homo sapiens and non-human animals is a common core awareness that is nevertheless expressed differently for each kind of organism at divergent levels overseeing management of disparate needs and actions, realized through behaviour in relation to concrete variations of the external environment. The dimension of "self-consciousness" is evolved, step by step, in phylogenetic progression according to an admirable order justifying the survival of each unique life form with respect to the particular tasks which it has to perform.

Today we are equipped with many high-end tools in our attempts to understand all the steps in the evolution of consciousness, but it is through intuition that we will achieve, simply, an adequate interpretation of consciousness itself, that most complex and extraordinary gift. Pending such 'intuition," some members of the QPP group have decided to submit to classical experimental testing those insights that each contributor has independently adduced through theoretical inquiry, that is, through the construction of an empirical map laying bare the most germane trans neuronal, neuronal, and sub neuronal molecular changes with an eye toward the possibility of inducing and measuring changes in membrane viscosity correlated with in vivo manifestation of mood disorders. As far as we know this will be the first time that such micro-molecular events are to be tied rigorously to molar cognitive phenomena.

The resulting experimental data may offer an enduring empirical anchor in contradistinction to the intersubjective vagaries that have afflicted those various psychiatric disease nosologies, most recently DSM V, issuing from the hollow consensus of committees and cultural contextual fashion. If the experimental program planned by the QPP group succeeds, the goal of psychodiagnostic validity, heretofore sacrificed by DSM to mere inter-rater "reliability", may at last be achieved.

5. A working hypothesis: Quantum Neuron Molecular Mapping (Q-NeMoMa) project

5.1. Numbers and figures of the experimental background

There is full knowledge that each mood disorder will manifest with different states of consciousness [17, 18].

Our platelets results, in their correspondence with the strong diagnostic power between Bipolar Disorder and Major Depression, and in the similarity found between human and animals, give the possibility to investigate the different neuronal genetic expression, and the possible inherited errors in neurons. The working hypothesis should provide neurons, from different animal origin, which are known to have molecular characteristics similar to those of man with mood disorders. Table 1 and Figure 10 [19-24].

This could allow understanding, first, the different gene expression according to the different psychiatric disorders studied; second, culturing the neurons belonging to the different animals, it will be possible to arrange modifications of the cell membrane viscosity. In agreement with the assumption, the path described can reasonably lead to the possibility to artificially create models of membrane viscosity corresponding to changes of the psychopathological phenomenon with the ability to achieve the set of molecular evaluations necessary for the understanding of the modifications of the interactome (*the whole [array of] molecular interactions that take place in an organism and allow the cascade of regulatory molecules including the mechanism of action of enzymes and metabolic reactions)*.

The Q-NeMoMa project, practically, wants to investigate the molecular modifications of the neuron according to different modifications of the viscosity of the neuronal membrane.

Some of the most important world experts have come together to identify the experimental procedures to be carried out.

Fatty Acids (% mean values)					
Animals	**Cases**	**Palmitic Acid**	**Linoleic Acid**	**Arachidonic Acid**	**B2 Index**
Sheep	4 pool of 3	19.91	8.22	4.73	3. 980
Bovine	4 pool of 3	18.37	26.72	6.77	2.937
Cat	4 pool of 3	17.45	27.75	9.54	2.240
Horse	4 pool of 3	14.8	23.17	6.46	2.173
Donkey	8 pool of 3	14.39	19.68	6.34	2.154
Guinea pig	Literature	17.4	12.4	14.6	1.675
Rat	Literature	24.40	9.5	20	2.567
Pig	80	26.09	8.78	14.12	3.957
German Shepherd	7	18.9	21.5	20.39	0.936
	1	15.13	20.65	22.00	-0.240
Alaskan Malamute	5	18.25	19	21.2	0.688
	1	16.7	17.89	23.93	-0.120
Humans (normal)	60	20.68	19.41	14.06	2.445
Depression 1 (MD+ BD)	84	17.92	16.71	19.03	1.002
Depression 2 (MD)	41	17.22	9.34	26.81	-0.310
Bipolar	67	19.75	8.65	23.79	0.819
Ischemia 1	50	23.32	10.51	15.17	3.072
Ischemia 2	87	19.59	4.74	12.72	2.658
Young adults	45	18.16	21	14.71	1.690
Children	59	23.23	11.82	10.77	3.746

Table 1. Average Fatty Acids and B2 index of different animals and human beings.

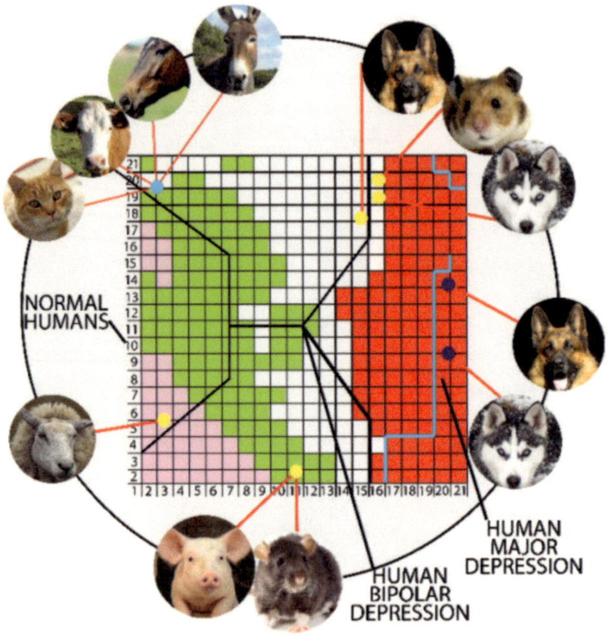

Figure 10. Distribution of animals and humans over the SOM.

From this important research will be possible to obtain data needed to assess whether, corrective actions for the improvement of the devastating conditions of all those who are suffering from mood disorders, will be possible.

A valuable help to the understanding of the neuron functioning can come from quantum molecular computation, by being able to interpret the neuron modifications, in the occurrence of the most important mood disorders such as Major Depression and Bipolar Disorder.

The suggested path could start from the largest scale: the cell membrane.

Five parallel approaches should be addressed, working one with the other, Figure 11:

1. Quantum chemical scale of neural signals by Bernroider [25, 26].
2. The Fatty Acid profile (Palmitic, Linoleic and Arachidonic Acids dynamics) of Cocchi and Tonello [2, 3, 7, 8, 9, 10, 11].
3. The role of lipid raft and G protein of Mark Rasenick [27-29].
4. Cytoskeleton modifications (Microtubules and Tubulins) studied by Tuszynski [30-33].
5. The exosomes studied by Francesco Cappello [34].

The complex dream we are running is to realize the molecular hypothesis of consciousness, designed in 2008 (private meeting in Bologna, Department of Veterinary Medical Sciences) by **Massimo Cocchi, Lucio Tonello, Mark Rasenick, Stuart Hameroff & Kary Mullis**. Figure 12.

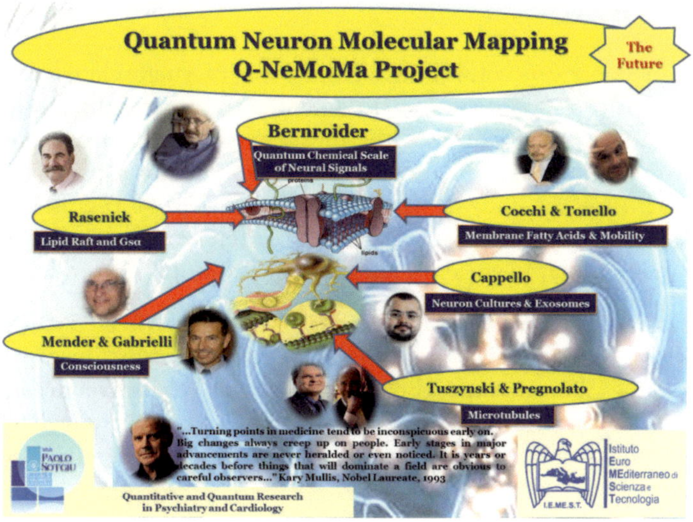

Figure 11. The steps of the project

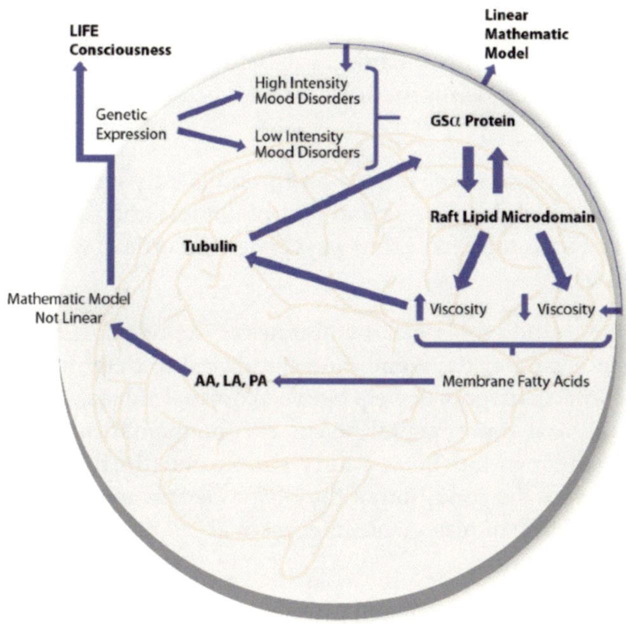

Figure 12. The consciousness molecular path

The whole path could be supervised by Gabrielli (Philosopher [8]) and Mender (Psychiatrist [35]), scientists of rigorous intellectual skills with a profound vision of the theoretical and conceptual aspects of psychopathology and quantum consciousness.

5.2. Final theoretical issues

Although there is evidence of a continuing effort by the international psychiatric community to refine the diagnosis of mood disorders, to date, the traditional diagnostic criteria are not enough sensitive in identifying patients with Bipolar Disorder (BD) from those suffering from Major Depression (MD), in the first diagnosis. Diagnosis remains mostly late and treatments, that may improve symptoms and quality of life, continue to be preceded by interventions which, in addition to not providing adequate relief, often worsen the BD course, increasing the likelihood of inducing rapid cycles or suicidal behaviour [50, 51].

Differential diagnosis of BD symptoms from other diagnoses has been documented as difficult [52-55]. Diagnosing BD from MD, psychosis, borderline personality disorders, obsessive-compulsive disorder, etc.) or neuropsychological disorders (cognitive impairment, dementia, etc.) or neuropsychological disorders, has presented challenges. Moreover, manifestations are highly variable not only from patient to patient, but also in the same subject at different stages of the clinical course and in later life.

To overcome this impasse various strategies have been identified and more sensitive and specific assessment tools have been searched for discriminating the BD condition and overcome the delay of an accurate diagnosis has been particularly difficult with MD. The BRIDGE study indicated a first way to go [56], and highlight the strength of some variables such as: mania/hypomania developing during therapy with an antidepressant or other drug, mood lability developing during antidepressant therapy, 2 or more prior mood episodes, and positive family history of mania/hypomania. A debate is essential between the advocates of traditional diagnostic and therapeutic methods and advocates of emerging methods resulting from new discoveries.

Major depressive disorder and other related and nonrelated psychiatric conditions are still characterised and defined by descriptive and non-biological criteria, but it is hoped that we can adequately characterise this and other psychiatric disorders with the addition of new quantitative approaches.

Cocchi and Tonello have studied platelet membranes of depressed subjects, enlisting profiles of FAs as a possible measure of the membrane status and to determine whether fatty acids could provide indications of diagnostic help between normal subjects and subjects affected by mood disorders. In the first experimental phase, two mathematical tools were identified, a complex one (Self Organizing Map (SOM)) and a simple one (the B2 Index), which will prove valuable not only to define the condition of the Major Depression and Bipolar Disorder, but also to provide the possibility of reasonable inferences about the biological significance of the two molecular mood disorders.

We see the emergence, in summary, of some stringent theoretical and anthropological focuses:

- Firstly, the distinction between first-level ontological Depression and second-level ontological Depression.
- The first kind of depression, of an existential nature, expressed by the most varied cultural traditions, is rooted in our structural contingency, which in time and in becoming recognises the mark of its own finiteness (man as an anguished and depressed "natural" animal).

- Second-level ontological Depression, on the other hand, refers to MD, understood as a molecular, bio-existential niche, marked cogently by serotonin and fatty acids, with its own specific "emotive tonality" [57, 58].

- The centrality of the concept of situation, that is of man as a being located "here and now", starting from biological markers able to offer the diagnosis of MD and BD an extremely real substrate, to be faced on an empirical, therefore public base, to construct a DSM that is not pregnant with pseudo-phenomenology or ideology, but which looks at the *dasein* of consciousness in its true biomolecular flesh [57, 58].

- The continuity between biology and culture, which finds full confirmation in the phenomenon of depression, whose original (ontological) biological nature (serotonin, fatty acids) thoroughly intercepts the structural (ontological) precariousness of living that the great cultural narrations have rooted in flesh and blood (ontic) existence.

- The view of man as a synthesis of *Körper* and *Leib*: life experience is always rooted in biology, the phenomenological "losability" and "impossibility" [59] are also, and firstly, biological "losability" and "impossibility".

To remain in the distinction, often denied only in the intentions, between *erklären* (causal explanation) and *verstehen* (psychological comprehension) means ignoring the prolific acquisitions of complexity theory; above all it means remaining prisoner to the "myth of the sense", on whose basis life experiences, the phenomenological approaches, the philosophical articulations, on which an authentic interpretation of the psychopathology should depend, are hypostatised. The reification of the metaphors, the empty sentimentalism of the interior resonances, the veneration of the illness *as a fruitful production of alternative worlds, the dilution of the tragedy of depression in the imaginative vis of melancholy*: this is the most injurious product of pseudo-phenomenology.

Also biology produces sense, indeed it is the original meaning on which to graft other forms of meaning, of which philosophy is undoubtedly a strong interlocutor, but alongside other forms of knowledge (biochemistry, quantum physics, biomathematics, anthropology, sociology...), as an overall, heuristic synthesis, expressive of an autonomous, therefore "adult", approach to psychopathology [57, 58].

Lastly, it is necessary that the scientific community and the world of the clinical profession commit themselves increasingly so that psychiatry and psychology can constitute themselves as heuristic bio-analytical-existential knowledge, where the diagnosis is not placed under the Heideggerian "yoke of the idea" [60, 61] (classificatory ideology and diagnostic imperialism), but refers to convincing biological markers. In this context, comparison with the neurosciences appears inescapable, particularly in their quantistic standpoint [62-67], with all the implications, also of an ethical nature, that this involves [68-78].

In other words, to start from biology to move towards increasingly complex systems, able to integrate biochemical expressions, living and irreducible existential experiences, social and cultural contexts. Depression therefore needs to be inscribed within a horizon of unmythicised, polyvocal meaning where the biological, physiological, clinical, existential, psycho-social and anthropological aspects are set as objective a hermeneutic framework as possible.

This is all the more so at a time like the present, when a person is often appraised only on the basis of successes achieved, of objects flaunted, of products voraciously consumed, in the instant, of his social visibility, of relentless efficacy, of perfect adaptation of the "thinkable to the possible": all contexts where the genetic and biological psychopathologies of mood are disproportionately amplified [79-83].

If the mechanistic-reductionist cognitive approaches have been characterised by the metaphor of the "edifice", of the solid Cartesian rock, all the forms of knowledge founded on complexity theory have been characterised by the metaphor of the "network", of thinking in relationships, in a dynamic, fluid, open manner. In the field of mental illness, this means setting aside both the organicist paradigm and the pseudo-phenomenological, "sentimentalistic", and therefore ideological, paradigm, in order to have an integrated view of biological objectiveness and humanistic psychotherapy.

That is to say, an expression of diverse interrelated contributions from the various disciplines (psychiatry, psychology, biochemistry, anthropology, quantum physics, mathematics, philosophy).

The observer thus becomes a builder of models, a manager of complexity, giving treatment the character of a truly empathic relationship. This is all the more so where distressing pathologies are involved, such as Major Depression and Bipolar Disorder, "caput mortuum" of psychiatry, because the absence of cogent biological markers seriously compromises every form of therapy. Hence the identification of a biological platform (fatty acids of platelets) as a starting point for a correct classification of MD with respect to BD.

6. Conclusion

The identification of three platelet fatty acids (Palmitic Acid-PA, Linoleic Acid-LA and Arachidonic Acid-AA), in addition to allowing the identification of subjects affected by Mood Disorders, brought about some hypotheses which, over the time, have been proven by robust experimental data concerning also the concept of serotonin uptake on the basis of membrane viscosity. Platelets, considered cells with high affinity to neurons, have the same embryonic origin of brain and skin (ectoderm). Over the last thirty years, numerous and influential works have reported a similarity between platelet and neuron's serotonin concentration, mainly in MD and BD.

This evidence, together with the possibility of classifying the two main mood disorders (Major Depression, MD and Bipolar Disorder, BD), led to some considerations on the molecular uniqueness of MD, generally understood as a phenomenon affecting only human beings, and, more precisely, just part of them. The identification of the characteristics that distinguish MD subjects from BD ones occurs through the different position on the SOM of the triplet of fatty acids, above mentioned, detected for each subject, and through the B2 chemical index.

In this context, we can trace the human states between normality, BD and MD, the latter considered as a bio-molecular and existential niche. Looking into the undeniable distinction made between depressed and bipolar subjects thanks to the neural network (SOM) and the chemical index (B2) for indirect assessment of platelet membrane viscosity, we asked ourselves the question of whether the molecular characteristics of subjects with MD were completely different from those of all other living beings both humans or animals. In the light of the experimental data, humans can have either positive or negative values of the B2 index. Those humans having positive values of B2 are normal (N), bipolar (B) and psychotic (P) people. On the contrary, major depressed subjects (MD) have negative B2 values. On the basis of our hypothesis, MD would be, at this point, the real disease, among all Mood Disorders, with specific molecular features and expressions of consciousness, according to the concept of Symmetry Breaking. The use of biochemistry, non-linear mathematics, and human-animal comparison leads to some reflections that are not only really close to a cultural and biological interpretation of mood disorders, but also pave the way for diagnostic perspectives and predictive interpretation patterns of the disease known as "Mood Disorder".

Appendix — From biology to the anthropology of treatment

It is undeniable that the depth of the *ens sofferens*, to be approached with a curative word [36], cannot be traced only to *disease*—pathology or biomedical classification—but also to *illness*—experience of the malaise as lived—and *sickness*—the social determination of the condition [37-39].

For all societies, illness is an event to be interpreted; it is not just a biological fact but also a cultural one. Basically, illness is representation, interpretation, of a portion or of all reality by individuals in a certain social context. The medical description of the human body and the illness always refer back to culturally peculiar meanings.

In strict terms, we could say that the pain articulates its meanings in suffering, which is a restless reflection on the ineluctable, and nevertheless unexpected, occurrence of the illness.

It can be understood, then, that only medical practice that does not limit itself to the biomedical dimension, which is in any case an indisputable point of departure, but is able to meet with the suffering person in his or her intimacy, can ensure, if not salvation against the perverse myth of recovery, at least dignity of the treatment as a profound ethical and existential relationship.

Hence the opening towards intimacy as a dual construction of meaning, the planning of significances contributed to both by the patient, in trusting abandon, and by the physician, as a warm, experienced responsibility. The intimacy is inhabited by the treatment (sphere of essential meanings of living) and not by anxiety (sphere of intra-worldly commerces), precisely because the experience of illness is experience of a relational wound which destructures the biological and existential narration of the subject:

- Wounded in relation to his own body, in *Der Zauberberg* Thomas Mann says that "illness makes men more corporeal, it makes them all body" [40], from which, on the one hand, we wish to distance ourselves as it is a sign of the precariousness of existence, while on the other hand we want to master it as we never did when in good health, because the anguish of feeling expropriated by it, not only by the almost fleshly possibility of death, but also by its medical visibility (the anguish—increasingly flaunted, for that matter, and reiterated like a mantra—of the reification of the medical approach), makes us feel the full weight of our vulnerability;

- A relational wound with respect to everyday life, whose narrative laceration provokes at first dismay, then a steady eclipse towards an indeterminate space-time, which for this reason is anguishing and inhospitable, and requires an approach that is not simply clinical but, precisely, one of intimacy, which, as a profound expression of empathy, configures itself as discretion, the word held back, the gesture experienced, total attention for the suffering countenance in a mutual exchange of meanings.

In the case of mental illnesses, then, the social, cultural weight takes on an almost transcendent value, due to the often blurred correspondences between classification and natural object, between nosology and effective reality of the illness.

When all this is recognised, what remains, excluding the interpretations of meaning, is the structural necessity, when coping with the illness—in our case the psychopathology—of starting from the biological fact, from objective biological markers, without which experience, biographical narrations, cultural rootings would be empty just as biology, on the other hand, would be blind.

The meaning is not only the prerogative of philosophy, which certainly remains a strong interlocutor, but also of biology and biochemistry: the corporeal meaning/significance of a pathological event.

Certainly, a medicine limited to the biological fact cannot be extensive and ostensive of the illness. At the same time, however, no one should doubt—as however happens unfailingly—that starting from a bio-medical platform can *ipso facto* reduces the physician to a pure functionary of the body and of the pathology connected with it.

Ultimately, we need to remind the alleged monolithic custodians of the thought of Husserl, Heidegger, Jaspers, Minkowski and Binswanger that rooting the pathology in biology does not mean expropriating the sick person of his illness and making the physician a mere functionary of the organism, an all-out pathologist who ignores biographies, experiences, corporeal dynamics and relational ontologies.

If anything, biographies, relationships, cultural expressions can be preserved in all their dignity, once their genuine biological matrix has been determined.

On the other hand, we risk only metaphysical hypostatisation and, therefore, a treatment rooted in an immobile metempirical "elsewhere".

A fruitful heuristic synthesis between *Körper* and *Leib*, *erklären* and *verstehen* [41], as a true commitment to healing the "living flesh" (*Fr. chair*) [42], is possible only if we start from the biological roots of our *being in the world* (*in der Welt sein*).

Heidegger's figures of omnipotence *(Allmacht)* and impotence *(Ohnmacht)* [43], can be taken up and re-elaborated in a synthesis between the naturalistic, classificatory power of science and the ever-open possibility of phenomenology, avoiding the same Heidegger's anti-technicist derailments and the exacerbated revivals of Binswanger's phenomenology as a mere therapeutic praxis, without theoretical rigour.

Like psychiatry and psychology—bio-reductionist and limited to the illness—so too "antipsychiatry", which reduces the illness only to a social construction, process of control, of exclusion/inclusion managed by bio-power [44-48], unable to recognise the productivity/creativity of schizophrenic processes [49], ends up by prejudicing the genuine dynamics of the treatment.

Hence the need for an objective biological reference able to act as a cogent platform in the treatment relationship, with reference to two distressing psychopathologies: MD and BD [1].

Appendix: Linoleic acid secrets

Linoleic acid, a regulator of the fine tuning?

On why you can read in the platelet, what happens in the neuron in the case of mood disorders, we are not, up to now, able to give complete answer.

It will help in trying to understand the phenomenon, the configuration of the level curves of the various fatty acids made in the SOM (Figure 13). The values of C20: 4 (Arachidonic Acid) stored in each artificial neuron ADAM were interpolated and distributed all over the map, depending on the distance weighted of minimum squares. A graphic profile, therefore, has been made and expressed in a 2D plane (Figure 14).

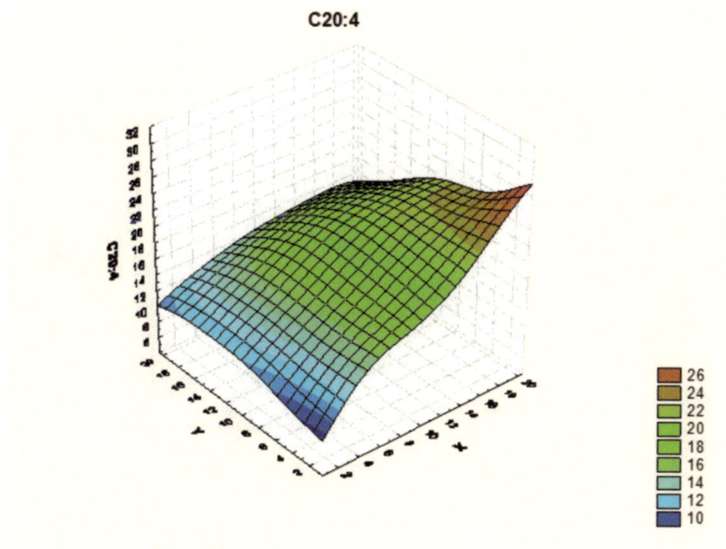

Figure 13. The value of C20: 4 stored in each artificial neuron of ADAM

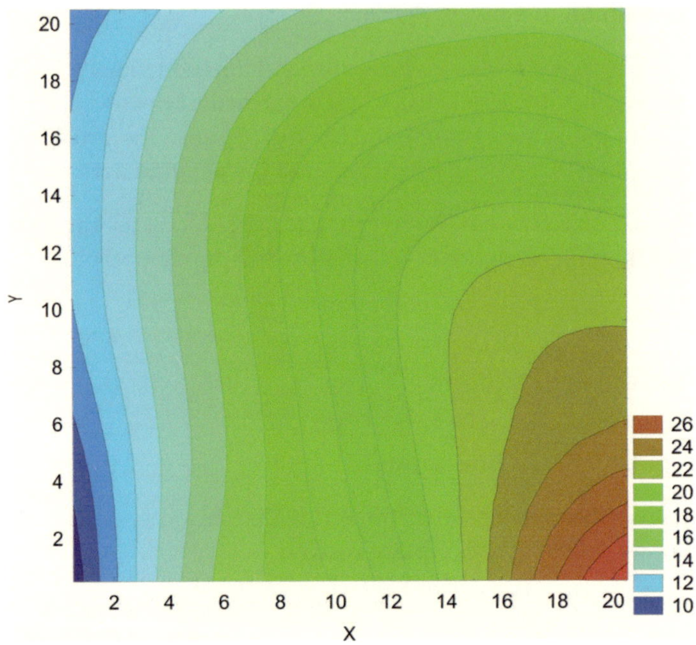

Figure 14. The level curves were made and expressed in a two-dimensional

Following the same procedure we have identified the maximum and minimum levels of the other fatty acids (Figure 15).

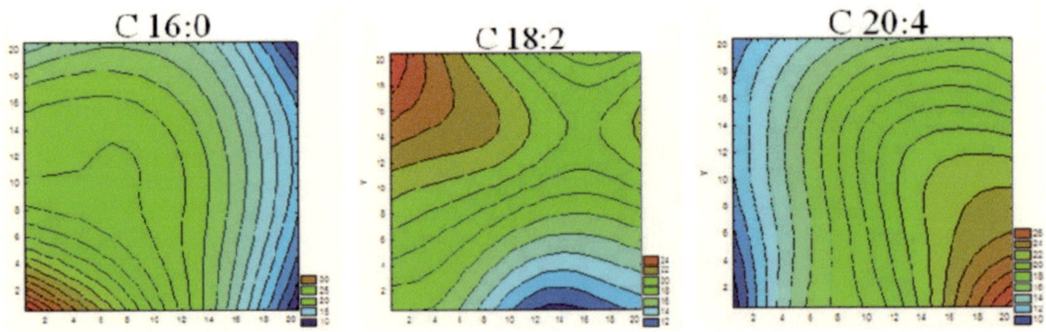

Figure 15. Minimum level (blue) and maximum (brown) of the fatty acids identified by the SOM

If we plot all three fatty acids expressed as index B2 we obtain Figure 16.

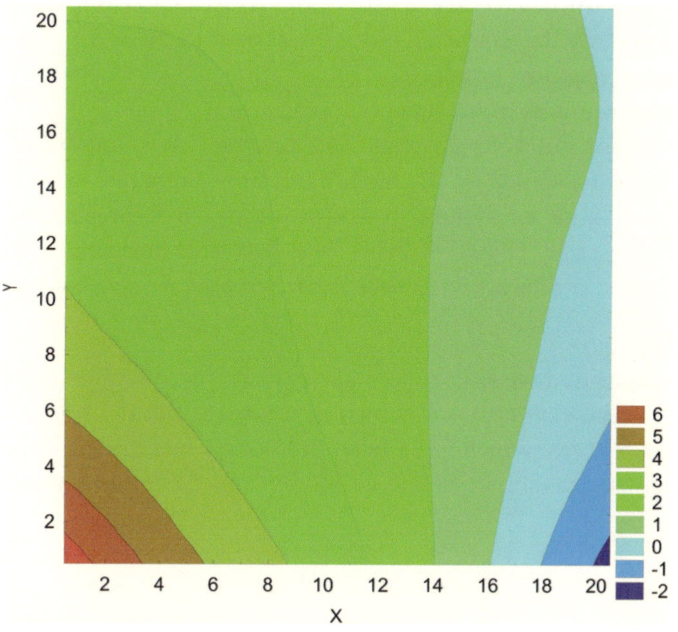

Figure 16. B2 index distribution over the SOM

The index B2, seems to be a good predictor of the macro-areas (Figure 17)

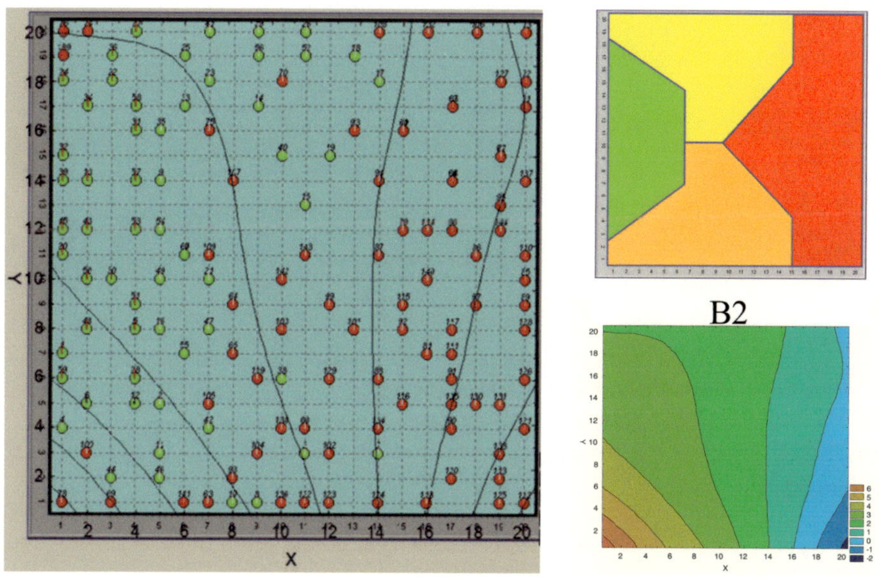

Figure 17. Level curves of the B2 Index over the SOM

A glimmer of light appears when we realize that only with the SOM and the B2 index, together, we can accurately identify the characteristic of the subject (SOM and B2 do not know of each other but converge on the same target). This observation becomes necessary when more than one subject, with the same B2 index have two different positions in the SOM, i.e., the same index can classify subjects in different areas. This finding is very important because it means that, in addition to the reasoning on the mobility of the membrane, there is another element of conditioning, and since everything revolves around the three fatty acids previously mentioned, must necessarily be one of them, in its concentration, that makes the difference and can affect the mood profile of the subject. Even in these cases we have accurate diagnostic findings.

For a variety of reasons that we will try to make explainable, attention is especially drawn to the linoleic acid, an essential fatty acid which is not manufactured by the human or animal organism. The level curves, which have been previously mentioned, show as the absolute minimum of linoleic acid, as shown below, corresponds to the minimum point of the B2 index (Figure 18):

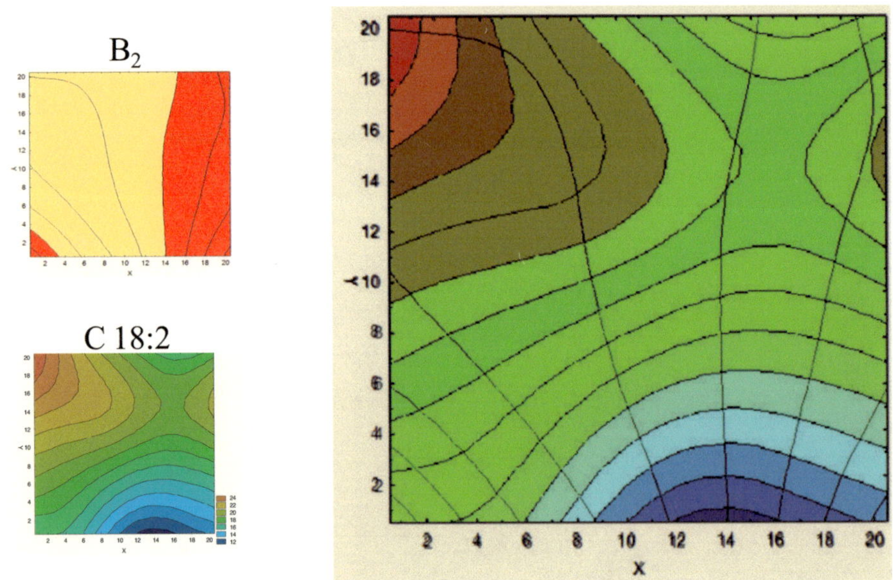

Figure 18. B2 index and linoleic acid distribution over the SOM

The fan produced by the SOM, from right to left, shows that the B2 is in progression from-2.64 To 8.23 (Figure 19).

The apparently healthy subjects are characterized by a mean value of B2 equal to 2.80.

This value is the midpoint between the extremes-2.64 (absolute minimum given by the map) and 8.23 (absolute maximum expressed by the map).

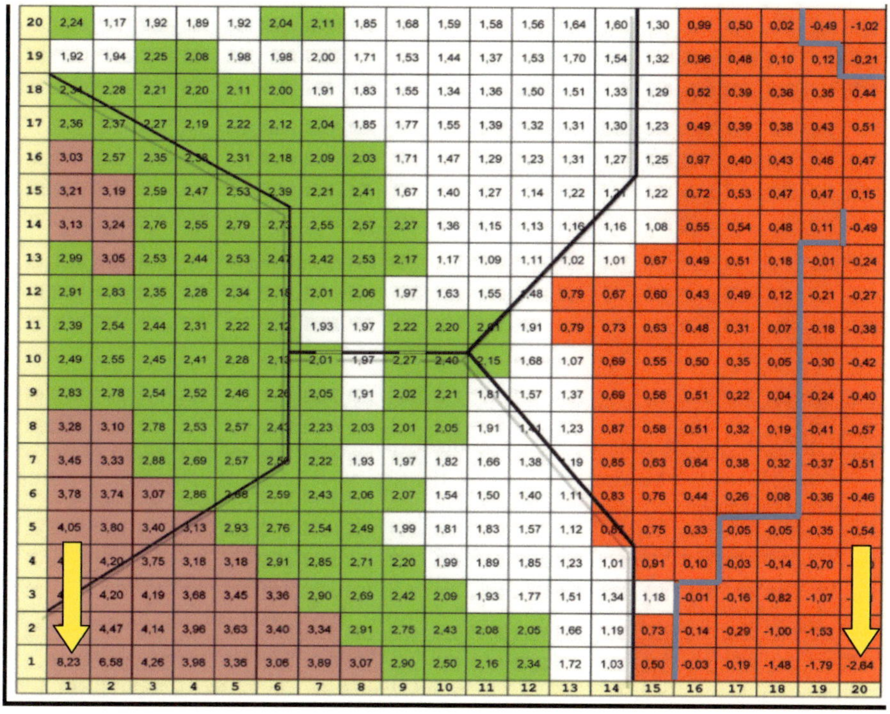

20	2,24	1,17	1,92	1,89	1,92	2,04	2,11	1,85	1,68	1,59	1,58	1,56	1,64	1,60	1,30	0,99	0,50	0,02	-0,49	-1,02
19	1,92	1,94	2,25	2,08	1,98	1,98	2,00	1,71	1,53	1,44	1,37	1,53	1,70	1,54	1,32	0,96	0,48	0,10	0,12	-0,21
18	[illegible]	2,28	2,21	2,20	2,11	2,00	1,91	1,83	1,55	1,34	1,36	1,50	1,51	1,33	1,29	0,52	0,39	0,36	0,35	0,44
17	2,36	2,37	2,27	2,19	2,22	2,12	2,04	1,85	1,77	1,55	1,39	1,32	1,31	1,30	1,23	0,49	0,39	0,38	0,43	0,51
16	3,03	2,57	2,35	[illegible]	2,31	2,18	2,09	2,03	1,71	1,47	1,29	1,23	1,31	1,27	1,25	0,97	0,40	0,43	0,46	0,47
15	3,21	3,19	2,59	2,47	2,53	2,39	2,21	2,41	1,67	1,40	1,27	1,14	1,22	[illegible]	1,22	0,72	0,53	0,47	0,47	0,15
14	3,13	3,24	2,76	2,55	2,79	[illegible]	2,55	2,57	2,27	1,36	1,15	1,13	1,16	1,16	1,08	0,55	0,54	0,48	0,11	-0,49
13	2,99	3,05	2,53	2,44	2,53	[illegible]	2,42	2,53	2,17	1,17	1,09	1,11	1,02	1,01	0,67	0,49	0,51	0,18	-0,01	-0,24
12	2,91	2,83	2,35	2,28	2,34	[illegible]	2,01	2,06	1,97	1,63	1,55	[illegible]	0,79	0,67	0,60	0,43	0,49	0,12	-0,21	-0,27
11	2,39	2,54	2,44	2,31	2,22	[illegible]	1,93	1,97	2,22	2,20	[illegible]	1,91	0,79	0,73	0,63	0,48	0,31	0,07	-0,18	-0,38
10	2,49	2,55	2,45	2,41	2,28	[illegible]	2,01	1,97	2,27	2,40	2,15	1,68	1,07	0,69	0,55	0,50	0,35	0,05	-0,30	-0,42
9	2,83	2,78	2,54	2,52	2,46	[illegible]	2,05	1,91	2,02	2,21	1,81	1,57	1,37	0,69	0,56	0,51	0,22	0,04	-0,24	-0,40
8	3,28	3,10	2,78	2,53	2,57	[illegible]	2,23	2,03	2,01	2,05	1,91	[illegible]	1,23	0,87	0,58	0,51	0,32	0,19	-0,41	-0,57
7	3,45	3,33	2,88	2,69	2,57	[illegible]	2,22	1,93	1,97	1,82	1,66	1,38	1,19	0,85	0,63	0,64	0,38	0,32	-0,37	-0,51
6	3,78	3,74	3,07	2,86	[illegible]	2,59	2,43	2,06	2,07	1,54	1,50	1,40	1,11	0,83	0,76	0,44	0,26	0,08	-0,36	-0,46
5	4,05	3,80	3,40	3,13	2,93	2,76	2,54	2,49	1,99	1,81	1,83	1,57	1,12	[illegible]	0,75	0,33	-0,05	-0,05	-0,35	-0,54
4	[illegible]	4,20	3,75	3,18	3,18	2,91	2,85	2,71	2,20	1,99	1,89	1,85	1,23	1,01	0,91	0,10	-0,03	-0,14	-0,70	[illegible]
3	[illegible]	4,20	4,19	3,68	3,45	3,36	2,90	2,69	2,42	2,09	1,93	1,77	1,51	1,34	1,18	-0,01	-0,16	-0,82	-1,07	[illegible]
2	[illegible]	4,47	4,14	3,96	3,63	3,40	3,34	2,91	2,75	2,43	2,08	2,05	1,66	1,19	0,73	-0,14	-0,29	-1,00	-1,53	[illegible]
1	8,23	6,58	4,26	3,98	3,36	3,06	3,89	3,07	2,90	2,50	2,16	2,34	1,72	1,03	0,50	-0,03	-0,19	-1,48	-1,79	-2,64
	1	2	3	4	5	6	7	8	9	10	11	12	13	14	15	16	17	18	19	20

Figure 19. Distribution of the B2 index over the SOM

A careful analysis of the mathematical formulation of the index B2 shows that it is governed almost entirely by the Arachidonic and Palmitic Acid. Intuitively this is deduced by expressing the average values of the two fatty acids of ADAM, using level curves, which appear qualitatively symmetrical (Figure 20).

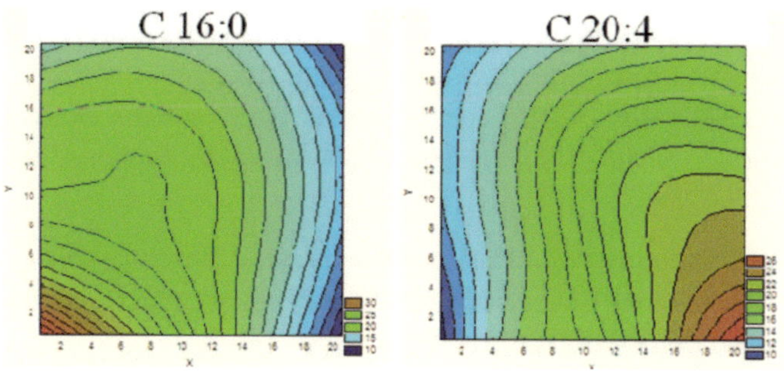

Figure 20. Demonstration of the symmetry of the distribution of palmitic acid (C16: 0) and arachidonic acid (C20: 4).

Are these two fatty acids that determine the macro area of a subject? Within a macro area is the intervention of linoleic acid that modulates with precision the position of the subjects. Maybe C18: 2 is the main actor in the "fine tuning"? (Figure 21).

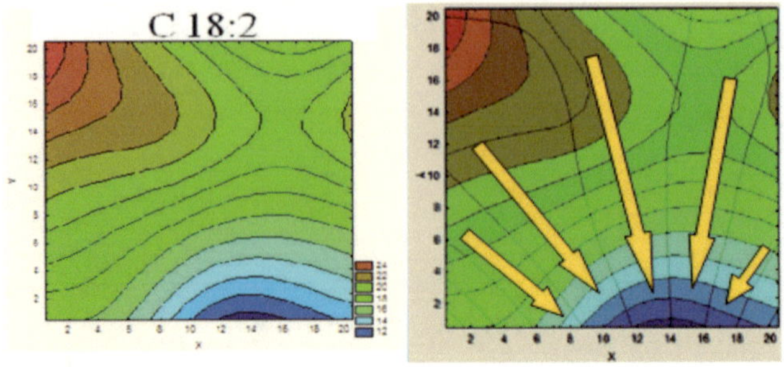

Figure 21. The distribution of linoleic acid on the map represents the core of the SOM

It is likely that in the circumstances that discriminate a healthy person from a pathological, the difference of linoleic determines, however, a modification of the biochemical factors involved and / or responsible for the biochemical and pathological determinism of the "Depression", of course declinable also on the basis of environmental / cultural influences. It should also be noted that the effects of linoleic acid on disease are relevant because of the configurations, equally balanced, of both, Arachidonic Acid and Palmitic Acid. Practically, referring to the positions of the subjects in the map ADAM, for values beyond a certain limit of Arachidonic Acid, the subject is surely depressed, as also happens for values beyond a certain limit Palmitic Acid. When B2 is in a neighbourhood of the normal value, it becomes decisive the percentage of linoleic acid. These configurations introduce, however, the concept of hyper saturation of the platelet in the case of Palmitic and that of hyper unsaturation in the case of Arachidonic. Another discriminant, between the two mentioned for the connotation of the disease, must be identified in Linoleic Acid which, in case of excess, restates the biochemical conditions for the development of the pathology. Certainly the network works beyond these operations, not yet known and interpreted probabilistically, of connection among the values that were administered.

A key point, that of linoleic acid, which requires the opening of a new chapter of considerations.

To better understand the reasoning on the data of linoleic acid, as well as his involvement in the molecular determinism of mood disorders, we must draw attention to some scientific findings that have linked, adversely, excess of linoleic acid, even for not very high concentrations, with some biological functions [84] and molecular interactions, i.e. the microtubules disruption [85].

A series of studies of cellular nutrition [86], on the effect of different amounts of phospholipids, extracted from various organs of calf (diencephalon, retina, cerebral cortex and heart), were made on chick embryo myocardial cultures.

From numerous tests, it was observed that the heart phospholipids, differently from the others, reduced, strongly, the migration speed of the cultures.

We did not understand at that time that the cardiac phospholipids, unlike the others (midbrain, retina and cerebral cortex), are very rich in linoleic acid, and that their addition, in addition to the amount naturally present, could be responsible for profound changes observed.

This effect could confirm once again the criticality of linoleic acid. Obviously, the consequences of the condition of excess will focus on the biological system in which the phenomenon occurs.

Even in case of reduction of linoleic acid may occur undesirable phenomena, as happens for example in the process of hibernation, in regulating the flow of calcium into the cardiac cell [87, 88].

The lipid structure of the brain as well as investigated [Cocchi and Noble, data not published, [89]] manifests the same characteristics in the extreme positions of the evolution of warm-blooded animal (from birds to humans) in the course of phylogeny [90], i.e. the level of Linoleic Acid is very low (about 0.3%).

It is possible to assume, reasonably, that while the manifestation of Mood Disorders is recognizable by the increase or decrease of specific fatty acids, Linoleic acid, even in its consolidated stability, could be the element capable of inducing, for small changes, amplifications of pathological brain responses.

Perhaps, within the concept of symmetry breaking and within the considerations on the linoleic acid, we can find answers to questions that the work done raises.

In particular we have, for a long time, faced the problem of how the set of three fatty acids could correspond with absolute precision to a condition of DM or DB. The perception that the set of identified molecular mechanisms might underlie the implications of quantum consciousness has been widely debated, finding aspects of great consistency in the molecular interactions involving membrane Gsα protein and cytoskeleton [9, 29, 28, 91, 92].

If we look at the map of the B2 index and the distance between the indexes (the expression of a molecular properties) we can realize how, in biology, the mathematical measure can express appreciable variations, in a numeric around, relatively close, as well as the positive and negative sign that characterizes respectively DB and DM, can never be interchangeable, consistent with the demonstration of the symmetry breaking between DM and DB.

Author details

Massimo Cocchi[1,2*], Lucio Tonello[1,3] and Fabio Gabrielli[1,3]

*Address all correspondence to: Massimo.cocchi@unibo.it

1 Institute "Paolo Sotgiu" Quantitative & Quantum Psychiatry & Cardiology, L.U.de.S. University, Lugano, Switzerland, Quartiere La Sguancia CH –Lugano-Pazzallo, Switzerland

2 Department of Medical Veterinary Sciences, University of Bologna, Via Tolara di Sopra 50, 40064 Ozzano dell'Emilia, Bologna, Italy

3 Faculty of Human Sciences, L.U.de.S. University, Lugano, Switzerland, Quartiere La Sguancia CH – 6912 Lugano-Pazzallo, Switzerland

References

[1] Bowden CL. (Strategies to reduce misdiagnosis of bipolar depression. Psychiatr Serv. 2001; 52: 51-55.

[2] Cocchi M, Tonello L, Tsaluchidu S, Puri BK. The use of artificial neural networks to study fatty acids in neuropsychiatric disorders. BMC Psychiatry 2008; 8(Suppl 1), S3.

[3] Cocchi M, Tonello L. Bio molecular considerations in Major Depression and Ischemic Cardiovascular Disease. Central Nervous System Agents in Medicinal Chemistry 2010; 10(2), 97-107.

[4] Kohonen T. Self-Organized formation of topologically correct feature maps. Biol. Cybern. 1982; 43, 59-69.

[5] Kohonen T. Self-Organizing Maps, 3rd ed. Springer: Berlin; 2001.

[6] Kohonen T, Kaski S, Somervuo P, Lagus K, Oja M, Paatero V. Self-organizing map. Neurocomputing 1998; 21,113-122.

[7] Cocchi M, Tonello L, Gabrielli F. Considerations on Blood Platelets: A Neuron's Mirror for Mood Disorders? Open Journal of Blood Diseases 2012; 2, 22-29.

[8] Cocchi M, Tonello L, Gabrielli F. Molecular uniqueness of major depression: biological remarks and theoretical implications. Journal of Consciousness Exploration & Research 2012; 3, 380-391.

[9] Cocchi M, Tonello L, Rasenick MM. Human depression: a new approach in quantitative psychiatry. Annals of General Psychiatry 2010; 9, 25.

[10] Benedetti S. et al. Platelet's Fatty Acids and Differential Diagnosis of Major Depression and,Bipolar Disorder through the Use of an Unsupervised Competitive-Learning Network Algorithm (SOM). Open Journal of Depression 2014; 3: 52-73.

[11] Cocchi M, Gabrielli F, Tonello L, Delogu M, Beghelli V, et al. Molecular Contiguity between Human and Animal Consciousness through Evolution: Some Considerations. J Phylogen Evolution Biol 2013; 1: 119.

[12] Pawlosky R, Barnes A, and Salem N Jr. Essential fatty acid metabolism in the feline: relationship between liver and brain production of long-chain polyunsaturated fatty acids. J Lipid Res. 1994; 35: 2032-2040.

[13] Pamplona R, Barja G, Portero-Otín M. Membrane Fatty Acid Unsaturation, Protection against Oxidative Stress, and Maximum Life Span A Homeoviscous-longevity Adaptation? Annals of the New York Academy of Science 2002; 959: 475-490.

[14] Dodman N. Obsessive Compulsive Disorder in Animals. Veterinary Practice News; 2011.

[15] Burns SD. OCD Equine. Brain Physic.com; 2011

[16] Cocchi M, Gabrielli F, Pessa E, Pregnolato M, Tonello L, Zizzi P. Major depression and Bipolar disorder: the concept of symmetry breaking. NeuroQuantology 2012; 10: 676-687.

[17] Flynn G, Alexander D, Harris A, Whitford T, Wong W, Galletly C, Silverstein S, Gordon E, Williams LM. Increased absolute magnitude of gamma synchrony in first-episode psychosis. Schizophr Res 2008; 105:262-271.

[18] Herrmann CS, Demiral T. Human EEG gamma oscillations in neuropsychiatric disorders. Clinical Neurophysiology 2005; 116: 2719–2733.

[19] Benedetti S, Bucciarelli S, Canestrari F, Catalani S, Colomba MS, Gregorini A, Mandolini S, Marconi V, Mastrogiacomo AR, Rasenick M, Silvestri R, Tagliamonte MC, Tonello L, Venanzini R, Cocchi M. Molecular Changes in Mood Disorders. Results of the Marche Region Special Project, NeuroQuantology 2012; 10: S1-28.

[20] Cocchi M, Sardi L, Tonello L, Martelli G. Do mood disorders play a role on pig welfare? Ital.J.Anim.Sci. 2009; 8: 691-70.

[21] Cocchi M, Tonello L, Gabrielli F, Pregnolato M, Pessa E. Quantum Human & Animal Consciousness: A Concept Embracing Philosophy, Quantitative Molecular Biology & Mathematics. Journal of Consciousness Exploration & Research 2011; 2: 547-574.

[22] Cocchi M, Tonello L, Gabrielli F. The animal side of "mood disorders", LAP Lambert Academic Publishing. Saarbrücken, Germany; 2012.

[23] Cocchi M, Gabrielli F, Tonello L, Delogu M, Beghelli V, et al. Molecular Contiguity between Human and Animal Consciousness through Evolution: Some Considerations. J Phylogen Evolution Biol 2013; 1: 119.

[24] Cocchi M, Tonello L. "Bio molecular considerations in Major Depression and Ischemic Cardiovascular Disease". Central Nervous System Agents in Medicinal Chemistry 2010; 9: 2-11.

[25] Bernroider G, Summhammer J. Can Quantum Entanglement Between Ion Transition States Effect Action Potential Initiation? Cognitive Computation 2012; 4: 29-37.

[26] Summhammer J, Salari V, Bernroider G. A Quantum-mechanical description of ion motion within the confining potentials of voltage gated ion channels. Journal of Integrative Neuroscience 2012; 11: 123-135.

[27] Cszysz AH, Rasenick MM. G-Protein Signaling, Lipid Rafts and the Possible Sites of Action for the Antidepressant Effects of N-3 Polyunsaturated Fatty Acids. CNS Neurol Disord Drug Targets 2013; Apr 4.

[28] Donati RJ, Dwivedi Y, Roberts RC, Conley RR, Pandey GN, Rasenick MM. Post-mortem Brain Tissue of Depressed Suicides Reveals Increased Gs Localization in Lipid Raft Domains Where it is Less Likely to Activate Adenylyl Cyclase. J. Neuroscience 2008; 28: 3042-3050.

[29] Allen J, Halvorsen R, Rasenick MM. Lipid Raft Microdomains and Neurotransmitter Signaling. NATURE Reviews Neuroscience 2007; 8: 128-140.

[30] Saha AA, Craddock TJA, Tuszynski JA. An investigation of the plausibility of stochastic resonance in tubulin dimers, Biosystems 2012; 107: 81-87.

[31] Craddock TJA, Tuszynski JA, Hameroff S. Cytoskeletal Signaling: Is Memory Encoded in Microtubule Lattices by CaMKII Phosphorylation? PLoS Comput Biol 2012; 8(3): e10024.21.

[32] Craddock TJA, Tuszynski JA. A critical assessment of the information processing capabilities of neuronal microtubules using coherent excitations. J Biol Phys. 2010; 36: 53-70.

[33] Woolf NJ, Craddock TJA, Friesen DE, Tuszynski JA.Neuropsychiatric Illness: A Case for Impaired Neuroplasticity and Possible Quantum Processing Derailment in Microtubules. NeuroQuantology 2010; 8: 13-28.

[34] Campanella C, Bucchieri F, Merendino AM, Fucarino A, Burgio G, Corona DF, Barbieri G, David S, Farina F, Zummo G, de Macario EC, Macario AJ, Cappello F. The odyssey of Hsp60 from tumor cells to other destinations includes plasma membrane-associated stages and Golgi and exosomal protein-trafficking modalities. PLoS One 2012; 7 (7): e42008.

[35] Mender D. From Quantum Wetware to Mental Illness: A Section Editor's First Interim Progress Report. First Interim Progress Report. NeuroQuantology 2010; 8: 115-119.

[36] Laín Entralgo P. The Therapy of the Word in Classical Antiquity. New Haven, CT: Yale University Press; 1970.

[37] Young A. The Anthropologies of Illness and Sickness. Annual Review of Anthropology 1982; 11, 257-285.

[38] Augé M., Herzlich, C. Le sens du mal. Anthropologie, histoire, sociologie de la maladie, Éditions des archives contemporaines, Paris; 1984.

[39] Good JB. Medicine, Rationality and Experience: An Anthropological Perspective. Cambridge: Cambridge University Press; 1994.

[40] Mann Th. The Magic Mountain. New York: Vintage Books; 1996.

[41] Husserl E. Cartesian Meditations: An Introduction to Phenomenology. Dordrecht: Kluwer Academic Publishers; 1977.

[42] Merleau-Ponty M. The Primacy of Perception: And Other Essays on Phenomenological Psychology, the Philosophy of Art, History and Politics. Evanston: Northwestern University Press; 1964.

[43] Binswanger L. Being in the World: Selected Papers of Ludwig Binswanger. New York: Basic Books; 1963.

[44] Goffman E. Asylums: Essays on the Social Situation of Mental Patients and Other Inmates. New York: Anchor Books; 1961.

[45] Foucault M. Madness and Civilization: A History of Insanity in the Age of Reason Tavistock. London; 1965.

[46] Foucault M. The Birth of the Clinic: An Archaeology of Medical Perception. New York: Vintage; 1994.

[47] Foucault M. Abnormal: Lectures at the Collège de France, 1974-1975. New York: Picador; 2004.

[48] Foucault M. Psychiatric Power: Lectures at the Collège de France, 1973-1974. New York: Picador; 2008.

[49] Deleuze, G., & Guattari. Anti-Oedipus, Continuum, London and New York, Vol. 1 of Capitalism and Schizophrenia 2004; 2 vols. 1972-1980.

[50] Phillips ML, Kupfer DJ. Bipolar Disorder Diagnosis: Challenges and Future Directions. Lancet 2013; 381, 1663-1671.

[51] Mazza M, Di Nicola M, Janiri L, Bria P. To Be or Not to Be a Bipolar Disorder Patient: Problems with Diagnosis. Journal of Nervous & Mental Disease 2013; 201, 435-437.

[52] Fountoulakis KN. Pharmaceutical Treatment of Acute Bipolar Depression. F1000 Medicine Report 2010; 2.

[53] Smith DJ, Griffiths KM, Hood K, Craddock N, Simpson SA. Unrecognised Bipolar Disorder in Primary Care Patients with Depression. British Journal of Psychiatry 2011; 199, 49-56.

[54] Suppes T, Datto C, Minkwitz M, Nordenhem A, Walker C, & Darko D. Effectiveness of the Extended Release Formulation of Quetiapine as Monotherapy for the Treatment of Acute Bipolar Depression. Journal of Affective Disorders2010; 121, 106-115.

[55] Sachs GS. Bipolar Mood Disorder: Practical Strategies for Acute and Maintenance Phase Treatment. Journal of Clinical Psychopharmacology 1996; 16, 32S-47S.

[56] Angst J, Azorin JM, Bowden CL, Perugi G, Vieta E, Gamma A, Young AH, BRIDGE Study Group. Prevalence and Characteristics of Undiagnosed Bipolar Disorders in Patients with a Major Depressive Episode: The BRIDGE Study. JAMA Psychiatry 2011; 68, 791-798.

[57] Cocchi M, Tonello L, Gabrielli F. Considerations on Blood Platelets: A Neuron's Mirror for Mood Disorders? Open Journal of Blood Diseases 2012; 2, 22-29.

[58] Cocchi M, Gabrielli F, Tonello L, Castaldo A. Music, Brain, Consciousness. Lugano: Ludes University Press; 2013.

[59] Minkowski E. Lived Time: Phenomenological and Psychopathological Studies. Evanston: Northwestern University Press; 1970.

[60] Heidegger M. Platons Lehre von der Wahrheit; 1931-32, 1942.

[61] Whitaker R. Anatomy of an epidemic: magic bullets, psychiatric drugs, and the astonishing rise of mental illness in America. New York: Crown Publishers; 2010.

[62] Pothos EM, Busemeyer, JR. Can Quantum Probability Provide a New Direction for Cognitive Modeling? Behavioral and Brain Sciences 2013; 36, 255-327.

[63] Woolf N, Craddock TJA, Friesen DE, Tuszynski JA. Impaired Neuroplasticity and Possible Quantum Processing Derailment in Microtubules. NeuroQuantology 2010; 1, 13-28.

[64] Pylkkänen P. Implications of Bohmian Quantum Ontology for Psychopathology. NeuroQuantology 2010; 8, 13-28.

[65] Zizzi P, Pregnolato M. Quantum Logic of the Unconscious and Schizophrenia Paradigms Quantum Logic of the Unconscious and Schizophrenia. NeuroQuantology 2012; 10, 566-579.

[66] Mender D. The Cunning of the Observable: Generalized Fourier Duality, the Hard Problem, and Natural Kinds of Psychopathology. Quantum Biosystems 2013; 5, 43-54.

[67] Cocchi M, Gabrielli F, Tonello L, Delogu M, Beghelli V, Mattioli M., Accorsi PA. Molecular Contiguity between Human and Animal Consciousness through Evolution: Some Considerations. Journal of Phylogenetics & Evolutionary Biology 2013; 1, 4, 2-8.

[68] Garland B. Neuroscience and the Law. New York: Dana Press; 2004.

[69] Furman R, Downey EP, Jackson RL. Exploring the Ethics of Treatments for Depression: The Ethics of Care Perspective. Smith College Studies in Social Work 2004; 74, 525-538.

[70] Gazzaniga M. The Ethical Brain. New York: Dana Press; 2005.

[71] Glannon W. Bioethics on the Brain. New York: Oxford University Press; 2006.

[72] Illes J, Sahakian BJ. Oxford HandDook of Neuroethics. Oxford: Oxford University Press; 2011.

[73] Levy N. (2007). Neuroethics. Challenges for the 21st Century. Cambridge: Cambridge University Press; 2007.

[74] Illes J, Sahakian BJ. Oxford HandDook of Neuroethics. Oxford: Oxford University Press; 2011.

[75] Benedikter R., Giordano J. Neurotechnology: New Frontiers for European Policy. Pan Euro Network Sci Tech 2012; 3, 204-207.

[76] Giordano J. Neuroimaging in Psychiatry: Approaching the Puzzle as a Piece of the Bigger Picture(s). AJOB-Neuroscience 2012; 3, 54-56.

[77] Kohls NB, DuRousseau D, Plischke H, Giordano J. Brain-Computer Interfacing (BCI): Potential Use and Ethico-Legal and Social Issues—Toward a Stance of Preparation and Steps Forward. Human Computer Interfacing International 2013; 10, Online.

[78] Monti MM. Ethics, Neuroimaging and Disorders of Consciousness: What Is the Question? AJOB Neuroscience 2013; 4, 1-2.

[79] Bauman Z. Consuming Life. Oxford: Polity Press; 2007.

[80] Bertman S. Hyperculture. The Human Cost of Speed, Praeger, Westport (CO)-London; 1998.

[81] Eriksen TH. Tyranny of the Moment: Fast and Slow Time in the Information Age, Pluto Press, London-Sterling; 2001.

[82] Gabrielli F. Philosophy and Psychiatry. The Violated Body in the Era of the Invisible Man. NeuroQuantology 2012; 10, 19-20.

[83] Cocchi M, Tonello L, Gabrielli F. Mood Psychopathologies: An Integrated Complexity-Based Interpretation. Psychology 2014; 5, 192-203.

[84] Marei WF, Wathes DC, Fouladi-Nashta AA. Impact of linoleic acid on bovine oocyte maturation and embryo development. Reproduction 2010; 139: 979–988.

[85] Namikoshi M, Suzuki S, Meguro S, et al. Inhibitors of Microtubule Assembly Produced by the Marine Fungus Strain TUF 98F139 Collected in Palau. Journal of Tokyo University of Fisheries 2002; 88: 1-6.

[86] Cocchi M, Marzona L, Pignatti C, Olivo OM. Effect of organ phospholipids on the growth of embryonal tissues cultured in vitro. Biochem. Exp. Biol. 1979; 15: 13-16.

[87] Giroud S, Frare C, Strijkstra A, Boerema A, Arnold W, et al. Membrane Phospholipid Fatty Acid Composition Regulates Cardiac SERCA Activity in a Hibernator, the Syrian Hamster (Mesocricetus auratus). PLoS ONE 2013; 8(5): e63111.

[88] Ruf T, Arnold W. Effects of polyunsaturated fatty acids on hibernation and torpor: a review and hypothesis, Am J Physiol Regul Integr Comp Physiol 2008; 294: R1044–R1052.

[89] Svennerholm L. Distribution and fatty acid composition of phosphoglycerides in normal human brain. Journal of Lipid Research 1968; 9.

[90] Cocchi M, Tonello L, Gabrielli F., Minuto C. Human and Animal Brain Phospholipids Fatty Acids, Evolution and Mood Disorders. Journal of Phylogenetics & Evolutionary Biology, in press.

[91] Cocchi M, Gabrielli F, Tonello L, Pregnolato M. The Interactome Hypothesis of Depression. NeuroQuantology 2010; 4: 603-613.

[92] Popova JS, Greene AK, Wang J, Rasenick MM. Phosphatidylinositol 4, 5-bisphosphate modifies tubulin participation in phospholipase Cβ1 signaling. J. Neurosci. 2002; 22: 1668-1678.

Chapter 4

The Effects of Psychotherapy on Brain Function — Major Depressive Disorder

Sang Won Jeon and Yong-Ku Kim

Additional information is available at the end of the chapter

http://dx.doi.org/10.5772/59405

1. Introduction

Functional neuroimaging technologies have played an important role in the observation and characterization of the change in the brain function during depression. Various treatment techniques for depression related to brain function changes have been investigated based on research on functional neuroimaging technologies. However, the development of such technologies has focused more on the effect of the pharmacotherapy neural mechanism than on the effect of psychotherapy on depression treatment. This has led to an unbalanced approach to depression treatment, which emphasizes pharmacotherapy more than psychotherapy even if the treatment cost and depression treatment efficacy of the former are similar to those of the latter.[1, 2] This imbalance was caused by the traditional perception of pharmacotherapy as a biological intervention and of psychotherapy as a psychosocial intervention. [3] However, the recent development of functional neuroimaging technologies proved that the changes in the affect/behavior/cognition after psychotherapy are caused by biological underpinning. Many recent studies on depression confirmed through neuroimaging that psychotherapy has neural correlations. These studies proved that psychotherapy is a biologically rigorous approach.[4]

The neurobiological change in, and the putative mechanism of, psychotherapy in depression have recently been highlighted.[5] These are very important mainly in two aspects. First, each psychiatric intervention should have a neuroscientific basis. Psychotherapy has been poorly supported scientifically until now. Its treatment efficacy on depression was empirical from history and experience. The effective development of pharmacotherapy requires a neuroscientific basis that enables estimation of the side effect or the concomitant effect of a drug. This is also the case for psychotherapy. Second, the relationship of psychotherapy and neurobiology would help provide neurobiological evidence of the efficacy of various therapies based on psychotherapy (e.g., cognitive behavior therapy (CBT) and neurofeedback). Scientific evidence of psychotherapy would promote its development.

The poor neurobiological evidence of psychotherapy was caused by difficulties in directly observing neural plastic changes in the human brain. Conventional methods of brain plasticity observation could be used only at the cellular level. The development of functional neuroimaging technologies enables non-invasive observation of plastic changes in the human brain. With these new technologies, training- and learning-related changes in the brain activation pattern could be observed [6] and applied in depression patients. In other words, changes in the brain function, especially in the neural mechanism, and changes in brain areas and circuits due to psychotherapy in depression, could be more closely observed. Moreover, the relationship between pharmacotherapy and psychotherapy could be investigated neurobiologically. By approaching the relationship between these two therapies neuroscientifically, an optimal therapy for depression can be chosen, and combined evidence of the efficacy of these two therapies can be prepared.[7, 8]

Most studies on functional imaging that analyzed the effect of psychotherapy on depression have used the nuclear medicine method. In such method, the brain metabolism and the blood flow on the pre- and post-treatment scans are compared using positron emission tomography (PET) or single photon emission computed tomography (SPECT). Functional magnetic resonance imaging (fMRI) is also very useful for measuring brain activation during depression treatment and the follow-up period with no exposure to radiation. However, fMRI is mainly useful for measuring the brain activation pattern during perceptual or cognitive tasks rather than for measuring the baseline brain metabolism. In measuring brain function changes using fMRI, two techniques are required: (1) one for provoking symptoms under MRI environments, and (2) the other for measuring the neural correlates of the psychopathology.[9]

In this chapter, the analysis of the relationship between psychotherapy and the brain function in neuroimaging studies for depression will be discussed. Systemic and critical reviews will be presented, and how psychotherapy changes the brain function of depression patients will be discussed.

2. Applications of neuroimaging for psychotherapy on depression

2.1. Understanding of consciousness and unconsciousness

It is common knowledge that unconsciousness is an integral part of the psychotherapy on depression. Recently, neuroimaging has made it possible to understand the unconscious process. According to several studies, subjects without depressive disorder have shown normal emotion and anxiety; however, the amygdala was activated after they have been exposed to a scary facial expression. Despite the fact that the subjects were unconscious of the stimulation, the amygdala, which is associated with anxiety, was activated when they were quickly exposed to a normal facial expression after transitioning from a scary facial expression. [10, 11] This indicates that anxiety is expressed through the unconscious process in patients with depressive disorder, which has been verified by neuroimaging. In another study, an amygdala hyperactivation, as well as a medial prefrontal cortex and an anterior cingulate cortex hypoactivation, occurred after exposing the patients with depressive disorder to the stimulation, and then examined through the use of the fMRI.[12] This result has been verified by fMRI, and it shows that the unconscious anxiety, which appears through the activation of the amygdala, cannot be consciously controlled in patients with depressive disorder.

2.2. Prediction of the risk factors on depression

Normal subjects, who have never experienced depression, were divided into two groups, namely, a group with a family history of depression and a group without a family history of depression. After processing the comparison, the volume of the hippocampus and prefrontal cortex was significantly reduced in the group with a family history of depression.[13, 14] This result explains some of the hereditary aspects of depression, thereby indicating the association of the neuroimaging findings with the risk factors on depression.[15, 16] In a study on functional neuroimaging, there is a remarkable mood-down-regulation effect due to the increased task in the normal group with a family history of depression than in the normal group without a family history of depression. Furthermore, the activation of the amygdala significantly increased in the normal group with a family history of depression as compared to the other group.[17] The activation of the putamen and insula was significantly lower in the normal group with a family history of depression than in the other group without a family history of depression in a reward expected situation, whereas the activation of the anterior cingulate cortex was significantly higher in the normal group with a family history of depression than in the other group in a punishment expected situation.[18] The aforementioned results are expected to explain some of the hereditary and risky aspects of complex depression. For this reason, it is interesting that the report on the abnormality of the limbic system and prefrontal cortex was observed in normal subjects without clinically significant depression.

2.3. Comprehensive development

Studies have been conducted in order to investigate the association of abnormal neuroimaging findings with trauma that has been experienced during childhood, and environmental and individual risk factors of depression. A previous study reported that the hippocampus volume was more reduced in patients who suffered from depression due to sexual harassment below 11 years of age than in patients who did not experience any sexual harassment at a young age. [19] In addition, the volume of the dorsomedial prefrontal cortex was more reduced in the group with child maltreatment than in the normal group.[20] The aforementioned studies were conducted in order to investigate how biological and psychological developmental factors, such as psychological trauma during childhood, progressed into pathophysiology of depression. However, the same psychopathology of depression is not always observed even if the subjects grew up in the same environment with the same genetic background.[21] Personal coping strategy, personality, and resilience contribute to the occurrence of depression, which makes the pathophysiology of depression more complex.[22-24] It was reported that the right amygdala- dorsomedial prefrontal cortex connectivity increased and the left amygdala-anterior cingulate cortex connectivity decreased in the normal group after they were exposed to negative emotions, such as fear and fury, as the severity of neuroticism increased.[25] In another study, a high extraversion group and a high neurosis group were exposed to positive/ negative stimulations, which resulted in the increased activity of the amygdala in the high extraversion group after they were exposed to a positive stimulation, as opposed to the high neurosis group after they were exposed to a negative stimulation.[25] The aforementioned studies were conducted in order to investigate how personality factors induce mood disorder by affecting the cognition and process of emotion.

2.4. Depressive disorder subgroups

The activities of each area of the brain can be assessed by neuroimaging, which makes the subgrouping of depressive disorders possible, according to the pathogenesis of depressive disorder. Depression has various disease etiologies, but expressive symptoms of depression are similar. It is also possible to subgroup depressive patients by neuroimaging based on their depression etiology. This explains why a specific treatment is effective on a depressive patient, but other treatments are ineffective on the same patient. For example, when elderly patients with depression were subgrouped into the white matter microstructural abnormality by the diffusion tensor imaging, the remission rate of depression was low.[26] As a result, the brain process of these patients can be understood by neuroimaging, which will make the personalized treatment possible.[27] Furthermore, the treatment course can be monitored by neuroimaging.

2.5. Prediction of the prognosis of depression

The prognosis of depressive disorder can be predicted by neuroimaging. For example, the prognosis was reported to be better in depressive patients with hypermetabolic rostral anterior cingulate cortex than in those with hypometabolic rostral anterior cingulate cortex when the depressive patients were examined by PET and fMRI.[28] In addition, the activation degree of the prefrontal cortex and orbitofrontal cortex by PET may vary in depressive patients depending on their treatment response.[29, 30] It was reported that depressive patients with hypometabolic orbitofrontal cortex responded better to pharmacotherapy, whereas those with hypermetabolic orbitofrontal cortex responded better to psychotherapy. Accordingly, studies on the prediction of treatment outcome by neuroimaging prior to therapy had been continuously conducted [31, 32] in order to investigate how new treatments of depression, such as mindfulness-based cognitive therapy and acceptance and commitment therapy, change the brain process [33, 34] by which the neuroimaging study will be closer to the actual clinical practice.

3. Neuroimaging study on depressive disorder

Many clinical, genetic, and biochemical studies reported that depression is not the manifestation of a single risk factor such as the abnormality of a single neurotransmitter, but instead, it is the overall abnormality of various genetic, environmental, and developmental risk factors. According to recent neuroimaging studies, depression is considered as an outcome of the abnormality of various interactions and systems rather than the abnormality of a single area of the brain.[35]

The cortico-limbic circuit is currently gaining attention due to its ability of mediating stress response and playing an important role in the regulation of emotions. In early neuroimaging studies, which investigated the role of the limbic area in depression, the structural abnormalities of the limbic area, such as the atrophy or bilateral loss of the amygdala [36, 37] and the volume reduction of the hippocampus [38, 39] and caudate nucleus [40], were consistently reported in depressive patients. In addition, functional neuroimaging studies reported that the increased activation of the amygdala [41] and the reduced connectivity of the amygdala-cingulate [42] were observed when the subjects were exposed to negative stimulation such as a scary facial expression. The limbic area, including the amygdala, recognizes and classifies the emotional stimuli consciously or unconsciously, and mediates autonomic nervous system responses according to the emotion.[43, 44] As a result, a constant report regarding the abnormality of the limbic area in depressive patients seems to be reasonable. However, the activation of the limbic area was more reduced when an emotional stimulation was presented in a conscious level than in an unconscious level.[45, 46] In addition, the reduced activation of the limbic area and the removal of the negative emotion were observed in healthy subjects through cognitive emotional control strategy such as reappraisal.[47, 48] Based on the aforementioned results, the hyperactivation of the limbic area alone is insufficient to describe depression.

Accordingly, neuroimaging studies have been mainly conducted on the frontal cortex, which mediates the cognitive emotional processing. The prefrontal cortex, which is a complex structure that mediates emotional control, attention, reward system, and response inhibition, is important as it interacts with the anterior cingulated cortex and the limbic area.[49] In early structural neuroimaging studies, anatomical abnormalities, such as the reduced frontal cortex volume [50, 51] and the reduced number and size of frontal cortex neurons [52] were reported in depressive patients. The functional neuroimaging studies reported that the reduced activation, blood flow, and metabolism of the dorsolateral prefrontal cortex and ventromedial prefrontal cortex were observed in depressive patients regarding the conduct of emotional task.[53, 54] Furthermore, the failure of processing the emotional stimulation, which was caused by the failure of controlling the activation of the limbic area due to the reduction of prefrontal cortex activation, resulted in the clinical symptoms of depression, such as negative emotion, rumination [55-57] and decreased execution function.[58] Based on the aforementioned results, the failure of controlling the hyperactivated limbic area, which is attributable to the decreased function of the prefrontal cortex that controls the recognition and process of emotional stimuli, is likely to be the most important neuroimaging finding on depression.

4. Neurobiologic mechanism of psychotherapy on depressive disorder

Many studies were conducted in order to investigate the importance of the amygdala in relation to depressive disorder. Studies related to fear were also conducted, and the results showed that fear acquisition occurred in the amygdala when a specific stimulation was presented. The acquisition of fear does not necessarily require the neocortex; however, it is necessary to eliminate fear.[59] Psychotherapy, including behavioral treatment, is effective in controlling implicit memory acquisition in the amygdala and prefrontal area. Dealing with symbolic factors in dynamic psychotherapy will also control the activation of the amygdala by the temporal lobe and neocortex. However, the circuit from the cortex to the amygdala is weaker than the circuit from the amygdala to the cortex.[60] This is the reason why a longer duration is required for a new connectivity formation from the cortex to the amygdala, which serves as an evidence of long-term psychotherapy requirement. Psychotherapy includes the intentional memory for explicit memory, and promotes implicit memory by free association. The patient's amygdala is activated during free association. This means that the patients may play active roles in psychotherapy, to which emotional and cognitive associations related to the original trauma can be treated. The process of psychotherapy on depression is not a passive conditioned reflex, but instead, it is an active learning process.[61]

There are some methods of biologically measuring abnormal cognitive and emotional patterns during psychotherapy using repression or unconscious forgetting.[62] A previous study used a method that requests a subject to remember a certain word and then forget it.[63] After the subject is requested to forget the word, the increase in the prefrontal cortex (PFC) activation and the decrease in the hippocampus activation were observed during the recall tasks.[64] Other previous studies showed similar results. The anterior cingulate gyrus is known to play an important role in the attempt to forget unwanted memories. As such, the PFC and the anterior cingulate play an important role in the intentional forgetting of words or memories: i.e., the biological measurement of repression in psychodynamic psychotherapy.

Psychotherapy is related to characteristic neural circuitries, and different neural circuits are activated in different types of psychotherapy. It has been confirmed that different types of psychotherapy, including psychodynamic psychotherapy, cognitive therapy, and behavioral therapy, have different activation processes. For example, in psychodynamic psychotherapy, the therapist collects information on the patient based on his episodic memory. However, if the therapist uses the free association method, which focuses more on random memories than on the episodic memory, when these random memories are induced, the areas in the brain such as the frontal, parietal, and temporal cortexes are associated. On the other hand, the episodic memory is associated more with specific verbal areas such as Broca's area and the left frontal operculum.[65] As such, free association is a less censored process with the use of the extensive cortex network. The therapist investigates the symptoms and personality of the patient based on the extensive cortex activation during psychotherapy.

When cognitive behavioral therapy (CBT) is applied in depression patients, the patients are reminded of bad or sad memories, and they re-evaluate their negative memories. They re-interpret negative memories more positively during CBT, and their emotions before and after the re-interpretation are rated. Several studies have investigated the correlations among the re-interpretations of the negative emotions/emotions/brain activity patterns during CBT.[66] The patients felt better after they re-interpreted their negative memories during CBT, which was related to the increased activity of their dorsolateral and dorsomedial PFC and to the decreased activity of their amygdala and orbitofrontal cortex. These findings are the bases of the biological model of CBT. Negative emotions as a result of certain stimuli are produced in the limbic and ventral prefrontal structures, which are decreased or blocked by the dorsal prefrontal circuitry during CBT.[67]

Finally, behavioral therapy (BT) will be discussed. BT is related to desensitization of stimuli that cause anxiety or extinction of learned responses. The results of animal tests and human studies showed that the areas, including the ventral PFC and the amygdala, are related to desensitization and extinction responses.[68] The result of an animal test using mice showed that the lesions in the amygdala or the administration of drugs that inhibit the amygdala distracted fear conditioning.[69, 70] Also, the results of human studies that used functional neuroimaging showed that amygdala activation is related to conditioned fear response. These results mean that the ventral PFC is related to the extinction and retention of the conditioned fear response by inhibiting the amygdala.[71] As such, the important mechanism of extinction-based BT involves the strengthening of the PFC activation and the attenuation of the amygdala.

5. Variable methodology of the brain function for psychotherapy and depression

The variability of the limitations and results of studies on psychotherapy and neuroimaging for depression should be reviewed first before their relationship is investigated. Recently introduced technologies on neuroimaging and their rationales are variable, and there are different ways to evaluate the efficacy of psychotherapy. Manual therapy under a time-limited setting is used for psychotherapy, including BT, CBT, and interpersonal therapy (ITP). Even if the program designs of many types of psychotherapy are standardized, they are subjective, depending on their investigator, due to their manualized treatment design.[72] Previous studies showed that the therapeutic modality of some types of CBT seem more similar to that of BT, or vice versa. The inconsistency of the methodology is also caused by the number of therapists (one or many), the number of sessions performed, or the type of therapy (e.g., individual therapy or group therapy). As such, the aforementioned factors should be considered when reviewing the results of previous studies.

Neuroimaging also has various modalities, as does psychotherapy. Neuroimaging is designed to evaluate the glucose metabolism or the cerebral blood flow (CBF). Although the relationship between the glucose metabolism/CBF and the neuronal activity cannot be understood fully, the brain function in neuroimaging is described using terms such as 'metabolic activity' or 'hemodynamic activity'. Also, "brain activity" is a commonly used term for both metabolic activity and hemodynamic activity.[73] As mentioned, neuroimaging refers to various methodologies, including PET, SPECT, and fMRI. Each imaging technique has a different mechanism, with different image resolutions and application limitations.[74] There are other methods of separately measuring the regional brain activity, such as voxel-based techniques (e.g., SPM) and region of interest (ROI)-based techniques. Multiple comparison is difficult for ROI-based techniques, but such techniques present clear anatomical borders. For voxel-based techniques, much data can be extracted, and clear functional connectivity can be achieved.[74]

Even if different neuroimaging and psychotherapy methods are used, they have a common and duplicate study design. Most imaging studies scan brain images before and after psychotherapy and compare the results with those of healthy control subjects to evaluate the treatment efficacy. A functional neuroimaging study measured the brain activity while a patient was resting, and another study measured the brain activity while a patient was exposed to anxiety stimuli or was performing cognitive or affective tasks. Understanding the differences between various neuroimaging methods can lead to accurate understanding of brain activity data.

6. Imaging neural effects of psychotherapy on depression

Various studies have been conducted in the last few decades on the neurobiological mechanism in depression and on its relationship with psychotherapy.[75, 76] Aaron Beck, the founder of CBT for treating depression, has continuously claimed that studies are required on the relationship between the psychological mechanism and the biological mechanism.[77] The recent controversy on whether or not the biological mechanisms of antidepressant medication (ADM) and psychotherapy are different has led to many studies on neuroimaging. Many previous studies have proven that ADM and psychotherapy have different neurofunctional mediators and different mechanisms. Moreover, the studies found that the therapeutic efficacy of ADM differs from that of psychotherapy between mild and severe depression [78], as do their relapse rates.[75]

Relatively consistent study results were shown on the brain function of the depression patients. The results showed that the resting metabolism of the amygdala increased in the depression patients [79], and that the amygdala activation increased while aversive stimuli were expected.[80, 81] On the contrary, patients with depression showed decreased prefrontal cortex (PFC) resting metabolism.[82, 83] The investigators proved, using fMRI, that the amygdalar reactivity in the depression patients was elevated during the emotion-processing tasks and that the dorsolateral prefrontal cortex (DLPFC) activation decreased during the cognitive tasks. The investigators also found that the functional connectivity between the amygdala and the DLPFC decreased in the depression patients.[54] These findings are the basis of the theory that the brain function of depression patients shows increased bottom-up emotional reactivity and decreased top-down regulation of emotions. This is called the 'cortico-limbic dysregulation model'.[84] The emotion regulation of healthy people is also modulated by the bottom-up/top-down interaction. This theological model best explains the mechanism of the neurofunctional mode and the dysfunctional network in depression hypothesized by Beck.[77]

Various experiment methods were used to discover the pathophysiological mechanism related to the neurofunctional mode and the dysfunction of a neural network in depression, using neuroimaging. In the beginning, many studies observed the changes caused by pharmacotherapy. One such study was that of Anand et al. [85], in which sertraline was administered to 12 depression patients for six weeks. Ten of them showed depression improvement, as well as elevated functional cingulo-limbic connectivity and naturally decreased limbic reactivity when they were exposed to aversive stimuli. These changes in the depression patients after the pharmacotherapy showed that the patients had pathologic functional connectivity and that depression is a network disorder.

This concept of a network disorder was also confirmed in longitudinal studies that compared depression patients and a control group treated with fluoxetine. All the patients showed elevated activities of their DLPFC, ACC, premotor, parietal, posterior insular, and posterior cingulate cortices and decreased activities of their subcingulate, parahippocampus, and thalamus after fluoxetine treatment.[86] Additional changes in the subcortical and limbic activities were observed in the patients who showed good response to fluoxetine. The isolated increase in the activity in the cortical structure shown in the control group was considered a top-down mechanism which is regarded as a placebo effect.

The principle of psychotherapy for depression does not seem to follow this simple principle of a decrease in the bottom-up regulation and an increase in the top-down regulation, based on the results of previous studies. The activity change in the PFC also differs between the tonic and resting states, and from the event-related responses in psychotherapy and pharmacotherapy.[75] With the authors' insufficient understanding of event-related fMRI and of the theory of the stimulus-dependent neural mechanism, there is only empirical evidence of the improvement of the brain function with psychotherapy, but the neurobiological fundamentals of psychotherapy have not been proven yet.

7. Posttreatment neuroimaging changes in depression

7.1. CBT and depression

[Goldapple]

There was a functional neuroimaging study that used only CBT, without pharmacotherapy (Goldapple et al., 2004).[87] The neuroimages before and after the CBT were compared, and the patients were asked not to ruminate on one topic during the FDG-PET. A further study by the same investigators included a patient group that underwent pharmacotherapy using only paroxetine. Surprisingly, the group that underwent only CBT showed decreased metabolism in multiple frontal regions, including in the dorsolateral PFC. This result that showed decreased dorsolateral PFC activity from the baseline is contrary to the previous results that showed CBT-strengthened dorsolateral PFC activity. The authors of this study understood this decreased prefrontal metabolism as a therapeutic effect of CBT due to decreased active rethinking and reappraisal of emotional ideas. This interpretation is consistent with the study results of Paquette et al. that showed the decreased dorsal PFC of spider phobia patients when their treatment was successful.[88] On the other hand, this result is opposite that of the study of Ochsnet et al., which showed that the ability to reappraise effect-generating stimuli was enhanced after CBT in the depression patients.[66, 67, 89] In the study of Goldapple, the PFC metabolism was elevated after the pharmacotherapy in the depression patients who received only paroxetine. This proves that the biological therapeutic mechanisms of pharmacotherapy and CBT in depression are different even if they have similar degrees of efficacy.[87]

The study of Goldapple also measured the changes in the limbic and paralimbic activities after CBT treatment in depression patients apart from the change in the PFC. They found that different mechanisms were involved in the patients who underwent only CBT and in the patients who underwent only paroxetine treatment. The patients who underwent only CBT showed clearly elevated activities in their hippocampus, parahippocampus gyrus, and dorsal cingulate gyrus. On the other hand, the patients who received only paroxetine showed less elevated activity in their hippocampal and parahippocampal areas with decreased activity in their posterior cingulate and ventral subgenual cingulate. Goldapple claimed a modality-specific model depending on the therapeutic response of depression [87], based on their study results and the widely known functional and anatomic relationship theory.[90, 91] Anti-depressant agents produce a bottom-up effect by disengaging the ventral frontal and limbic region, whereas CBT produces a top-down effect by decreasing the cortical processing. These changes help process of personal emotions and environmental stimuli, and show a therapeutic effect on depression. The results and interpretation of Goldapple et al. are opposite those of the emotional and regulation model [66] of Ochsner et al. Ochsner understood that ventral frontal and limbic hyperactivity caused negative emotions. Seminowicz et al. [92] explained the difference between the theories of Goldapple and Ochsner according to the brain activation, depending on the treatment protocol and the underlying brain state. In other words, Goldapple et al. performed CBT on depression patients, whereas Ochsner used it on healthy control subjects. As such, future studies are required to investigate the change in the brain activity when a different type of psychotherapy apart from CBT is used for depression treatment.

[Kennedy]

Kennedy performed a study with the same design as that of Goldapple to analyze neuroimaging, after dividing the depression patients into two groups: the CBT group and the pharmacotherapy group treated by venlafaxine (Kennedy et al., 2007).[93] Kennedy hypothesized that the prefrontal activity would decrease due to top-down processing after the CBT, and that subcortical changes would appear due to bottom-up modulation after the administration of venlafaxine. The glucose metabolism of the patients was measured before the treatment, after the treatment, and 16 weeks after the completion of the treatment following the CBT and the venlafaxine administration. All the patients who responded to the treatment, regardless of the CBT or the venlafaxine administration, showed decreased activities in their orbitofrontal cortex and their left medial prefrontal cortex, and increased activity in their right occipito-temporal cortex. The patients who responded to the CBT showed increased activities in their posterior cingulate and thalamus, and decreased activity in their left inferior temporal cortex. On the other hand, the patients who responded to the pharmacotherapy (venlafaxine) showed increased activity in their left temporal cortex and decreased activity in their posterior cingulate. The authors of the study concluded that CBT is related to reciprocal modulation of cortical limbic interactions, which was consistent with the results of previous studies. They also presented venlafaxine as related to the cortical and striatal region, which was not known before.

[FU]

Fu reported the effect of CBT on depression using fMRI (Fu et al., 2004).[94] In their study, the fMRI scans of 16 depression patients were compared with those of healthy control subjects. When the 16 depression patients were exposed to a sad facial expression before the CBT, the reactivity of their right amygdala and hippocampus relatively increased compared to that of the healthy subjects. They also showed decreased ACC activation at the baseline, unlike the healthy subjects.[94] These differences between the groups no longer appeared after 16 CBT sessions. This means the stimulus-dependent reactivity in the amygdala decreased and the ACC activation increased in the depression patients after the CBT. However, it is uncertain if this result is meaningful, as the evidence of the event-related fMRI methodology is still insufficient and the stimulus-dependent neural mechanism is not fully understood yet. It has been understood that psychotherapy does not change certain areas but rather, extensive areas in the brain, and generally modifies brain dysfunction.[95] However, the results of the investigation of how psychotherapy changes the metabolic or functional activity have been inconsistent. In particular, the effect of psychotherapy on the PFC metabolism is more controversial. This may be attributed to the differences in the methodologies of the studies.

7.2. ITP and depression

[Martin]

Martin [96] measured the change in the CBF after IPT in depression patients. The neuroimages of the depression patients who underwent IPT were compared with those of the depression patients who underwent pharmacotherapy. The patients who underwent pharmacotherapy received 37.5 mg of venlafaxine daily. Twenty-eight depression patients were treated for six weeks, and their CBFs were measured using 99mTc-HMPAOSPECT. All of them had not been treated with any drug in the last six months before they participated in the study. The patients in the IPT group underwent IPT for six weeks with the same therapist, and those in the pharmacotherapy group were administered venlafaxine every second week with a 15-minute consultation. The patients in both groups showed clinical improvement with increased blood flow in their right basal ganglia. The patients in the IPT group showed additional changes - increased right posterior cingulate activity -which was similar to the result of the study of Goldapple.[87] Martin had the same opinion of Goldapple's result and considered insignificant the changes in the limbic and paralimbic recruitment of the brain function related to psychotherapy. However, he also observed the change in the brain function in a specific paralimbic region. A study is required to interpret this result.

[Brody]

Brody [97, 98] conducted a study with a design similar to that of Martin for 12 weeks. The study was conducted on 24 patients, divided into two groups: the IPT group and the paroxetine group. Then their PET neuroimages were compared. The study also included a healthy control group, unlike Martin's study. The patients chose which treatment they would receive. The results of the study showed that the symptoms of the patients who chose paroxetine treatment were milder than those of the patients who selected IPT, and showed greater improvement after the treatment. Even if there were differences between the groups, the results of the study showed that the dorsal and ventral prefrontal cortical metabolism decreased after the IPT. This result was consistent with that of the study of Goldapple.[87, 92] All the patients in both groups showed increased metabolism in their limbic and paralimbic regions, particularly in their right insula and left inferior temporal lobe. Unlike Goldapple's study, Brody's study reported that the patients treated with paroxetine showed decreased PFC activation.[97, 98]

Brody conducted a further study on 39 patients.[99] As in his previous study, the subjects were divided into two groups: the IPT group and the paroxetine group. Their neuroimages, obtained using PET, were analyzed. Unlike in Brody's previous study, however, the improvements in the specific mood symptom cluster and in the brain activity were compared. In all the patients, decreased ventral and dorsal frontal lobe metabolism was associated with reduction of symptoms such as anxiety/somatization, psychomotor retardation symptom, tension/anxiety, and fatigue. It was interesting that the improvement of the cognitive function was positively correlated with the degree of the dorsolateral PFC metabolism. This result differed from those of previous CBT studies. Some studies reported that the activity of the dorsolateral PFC was negatively correlated with the global depression score after the CBT. This difference was attributed to the reduction of over-thinking and rumination and the decrease in the dorsolateral PFC function with CBT, and the increase in the general cognitive ability and the strengthening of the dorsolateral PFC function with IPT. However, the difference should be verified with further repetitive studies. Brody's study had the following significant limitations: (1) the patients chose between IPT and paroxetine administration for the treatment type, which caused a difference between the groups, and (2) six symptom clusters, each with 12 ROIs (regions of interest), were used for the analysis without correction for multiple comparisons.

7.3. PDT (PsychoDynamicpsychoTherapy) and depression

The metabolic activities in the amygdala, hippocampus, and dorsal prefrontal cortex of the depression patients treated by PDT become similar to that of healthy people when the patients are exposed to attachment-related stimuli. The activation of the subgenual cingulate cortex has been known to decrease in depression patients after PDT, compared to healthy people [100], and this neural mechanism of PDT in relieving the symptoms of depression was proven through neuroimaging. Depression is believed to be a pathological state of the activity level of the brain circuit, and PDT is believed to normalize the brain circuit.

The designs of the studies on the neural mechanism of PDT through neuroimaging were similar to those of CBT or IPT, as previously stated. The change in the brain function was analyzed before and after PDT, and the results of the control and pharmacotherapy groups were compared.[101] However, the difference in the methodologies of the studies makes their comparison difficult. For example, Buchheim suggested that PDT was very effective in patients with subgenual ACC overactive depression [100], whereas CBT did not show any relevance. [102, 103] It was not completely understood if such result was because the neurobiological mechanisms of PDT and CBT were different, or if their study methodology was different due to their significantly different therapy methods. As such, standardization of the methodology is crucial in analyzing imaging data.

[Buchheim]

The study examining the effect of psychotherapy on brain function in patients with depression was conducted to investigate the effect of long-term psychodynamic intervention (Buchheim et al., 2012).[100] Sixteen, un-medicated outpatients with depression who underwent 15 months of psychodynamic therapy were scanned twice, before and after treatment, during the presentation of attachment-related scenes with neutral descriptions alternated with descriptions containing personal sentences previously extracted from an attachment interview. The results showed increased activation in the left anterior hippocampus/amygdala, subgenual cingulate, and medial prefrontal cortex in patients compared to healthy controls before treatment, and a reduction in the same areas after treatment. Furthermore, this normalisation of brain activity was positively correlated with general symptom improvement.

[Hirvonen, Karlsson]

Two other PET studies also focused on changes in neurotransmission implicated in the pathophysiology of depression following short psychodynamic psychotherapy (Hirvonen et al., 2010; Karlsson et al., 2010).[104, 105] Karlsson et al. (2010) for example compared the effect of psychotherapy and Fluoxetine on the density of serotonin 5-HT1A receptors, building on previous studies reporting a widespread decrease in the density of serotonin 5-HT1A receptors in the disease.[106-108] Specifically, patients were randomly assigned to either pharmacotherapy or psychotherapy, with the subsequent groups comprising 15 and 8 subjects respectively. Binding potential (BP) values, representing the ratio of specific and non-displaceable binding, were estimated using white matter as reference region (BP is a crucial measure in the PET studies to measure the density of "available" receptors), [109] Although both groups showed comparable symptom improvement post-treatment, when pre- and post-treatment values were compared, only those who underwent psychotherapy showed increased serotonin 5-HT1A binding in several cortical regions including dorsolateral prefrontal cortex, ventrolateral prefrontal cortex, ventral ACC, inferior temporal gyrus, insular cortex, and the angular gyrus.

In a subsequent study by the same group (Hirvonen et al., 2010), the effect of short psychodynamic psychotherapy and Fluoxetine was compared in patients with major depression with a specific focus on striatal and thalamic dopamine D2/3 receptors.[104] Results showed that although both treatments led to a significant improvement in symptomatology, no effects on D2/3 receptor availability in the ventral striatum or other subdivisions of the striatum were found. Moreover, only Fluoxetine increased thalamic D2/3 binding, but this increase was not correlated with clinical improvement. Thus, the study does not support the involvement of ventral dopaminergic neurotransmission in the antidepressant effects of Fluoxetine or psychodynamic psychotherapy. However, statistical power may have been hindered by the relatively small sample size.

7.4. BADT (Brief Behavioural Activation treatment for depression) and depression

Dichter et al. also carried out 2 fMRI studies with the same sample (2009, 2010). The main goal of the first study (Dichter et al., 2009) was to elucidate the neural correlates of Brief Behavioural Activation Treatment for Depression (BATD) during reward processing in depressed patients.[110] BATD is a structured and validated psychotherapy designed to increase engagement with rewarding behaviours and reduce avoidance behaviours.[111] Brain activity was measured in 12 subjects with depression and 15 healthy controls, while they were engaged in a Wheel of Fortune task (WOF), a two-choice decision-making task involving probabilistic monetary outcomes, before and after BATD.

Following the BATD, patients showed increased activity in structures that mediate responses to rewards, including the paracingulate gyrus during reward selection, the right caudate nucleus during reward anticipation, and the paracingulate and orbital frontal gyri during reward feedback. In contrast, the main aim of the second study by Dichter et al. (2010) was to identify baseline fMRI predictors of response to treatment in depression.[112] Before and after BATD, they scanned patients and matched controls while they were performing a task requiring cognitive control in both sad and neutral contexts. The results showed that following treatment, there was decreased activity in prefrontal structures including the paracingulate gyrus, the right orbital frontal cortex and the right frontal pole in response to stimuli presented within a sad context. In addition, pre-treatment activity in the paracingulate gyrus was identified as a significant predictor of symptomatic improvement following treatment.

8. Conclusion

In summary, despite the heterogeneity of the studies in terms of neuroimaging technique, psychotherapeutic approach and experimental paradigm employed, the majority of the above results suggest that psychological treatment of patients with depression results in a normalisation of the activation pattern in fronto-limbic circuitry. Regarding the relationship between the effects of psychotherapy and medication, their mechanisms seem divergent based on the results of two independent studies.[84] To explain these different mechanisms of action, it has been suggested that while psychotherapy may exert its effects top-down, targeting mainly frontal cortical regions and reducing dysfunctional thought processes, pharmacotherapy may produce bottom-up changes by disengaging ventral and limbic regions mediating attention to personally relevant emotional and environmental stimuli.[87, 92]

Neuroimaging techniques have significantly contributed to biological psychiatry research in the last decades, but the investigation of the mechanism of psychotherapy for depression has still been limited. The neural mechanism of psychotherapy is widely recognized, whereas its process cannot be clearly described yet. As psychotherapy holds a key position in psychiatry, it will be continuously investigated and researched. Therapy without a neuroscientific basis is only empirical, and the possibility of its development is low. Also, the application of such therapy is difficult, with limited potential to develop towards a new therapy.

Study	Psychotherapy (Number of subjects)	Pharmacotherapy (Number of subjects)	Controls	Neuroimaging Methods
Goldapple et al.(2004)	CBT (14)	Paroxetine (13)		PET
Kennedy et al.(2007)	CBT (12)	Venlafaxine (12)		PET
Fu et al.(2008)	CBT (16)		16	fMRI
Martin et al.(2001)	IPT (13)	Venlafaxine (15)		SPECT
Brody et al.(2001)	IPT (14)	Paroxetine (10)	16	PET
Buchheim et al.(2012)	15month PDT (16)		17	fMRI
Hirvonen et al.(2010)	Short term PDT (8)	Fluoxetine (14)		PET
Karlsson et al.(2010)	Short term PDT (8)	Fluoxetine (15)		PET
Dichter et al.(2009,2010)	BADT (12)		15	fMRI

Table 1. Studies on the effects of psychotherapy on brain function in depressive patients.

Acknowledgements

We wish to thank Yong Chon Park, Professor, Department of psychiatry, College of Medicine, Hanyang University Guri hospital, for his comments on this manuscript.

Author details

Sang Won Jeon and Yong-Ku Kim*

*Address all correspondence to: yongku@korea.ac.kr

Department of psychiatry, College of medicine, Korea University, Ansan Hospital, Jeokgeum-ro, Danwon-gu, Ansan-si, Gyeonggi-do, Republic of Korea

References

[1] Antonuccio DO, Danton WG, DeNelsky GY. Psychotherapy versus medication for depression: challenging the conventional wisdom with data. Professional Psychology: Research and Practice 1995;26:574–585.

[2] Goldman W, McCulloch J, Cuffel B, Zarin DA, Suarez A, Burns BJ. Outpatient utilization patterns of integrated and split psychotherapy and psychopharmacology for depression. Psychiatric Services 1998;49: 477–482.

[3] Westen D, Novotny CM, Thompson-Brenner H. The empirical status of empirically supported psychotherapies : assumptions, findings, and reporting in controlled clinical trials. Psychological Bulletin 2004;130: 631–663.

[4] Kandel ER. A new intellectual framework for psychiatry. American Journal of Psychiatry 1998;155:457–469.

[5] Kandel ER. Biology and the future of psychoanalysis: a new intellectual framework for psychiatry revisited. American Jornal of Psychiatry 1999;156:505–524.

[6] Linden DEJ. Cerebral mechanisms of learning revealed by functional neuroimaging in humans. In: Ku°hn R, Menzel R, Menzel W, Ratsch U, Richter MM, Stamatescu I-O (eds). Adaptivity and Learning – An Interdisciplinary Debate. Springer: Heidelberg; 2003. p 49–57.

[7] Brody AL, Saxena S, Schwartz JM, Stoessel PW, Maidment K, Phelps ME et al. FDG-PET predictors of response to behavioral therapy and pharmacotherapy in obsessive compulsive disorder. Psychiatry Research 1998;84:1–6.

[8] Hendler T, Goshen E, Tzila Zwas S, Sasson Y, Gal G, Zohar J. Brain reactivity to specific symptom provocation indicates prospective therapeutic outcome in OCD. Psychiatry Research 2003;124:87–103.

[9] Dierks T, Linden DEJ, Jandl M, Formisano E, Goebel R, Lanfermann H et al. Activation of Heschl's Gyrus during auditory hallucinations. Neuron 1999;22:615–621.

[10] Etkin A, Klemenhagen KC, Dudman JT, Rogan MT, Hen R, Kandel ER, et al. Individual differences in trait anxiety predict the response of the basolateral amygdala to unconsciously processed fearful faces. Neuron 2004;44:1043-1055.

[11] Ekman P, Sorenson ER, Friesen WV. Pan-cultural elements in facial displays of emotion. Science 1969;164:86-88.

[12] Hull AM. Neuroimaging findings in post-traumatic stress disorder. Systematic review. British Journal of Psychiatry 2002;181:102-110.

[13] Chen MC, Hamilton JP, Gotlib IH. Decreased hippocampal volume in healthy girls at risk of depression. Archives of General Psychiatry 2010;67:270-276.

[14] Baaré WF, Vinberg M, Knudsen GM, Paulson OB, Langkilde AR, Jernigan TL, et al. Hippocampal volume changes in healthy subjects at risk of unipolar depression. Journal of Psychiatric Research 2010;44:655-662.

[15] Amico F, Meisenzahl E, Koutsouleris N, Reiser M, Möller HJ, Frodl T. Structural MRI correlates for vulnerability and resilience to major depressive disorder. Journal of Psychiatry and Neuroscience 2011;36:15-22.

[16] Nolan CL, Moore GJ, Madden R, Farchione T, Bartoi M, Lorch E, et al. Prefrontal cortical volume in childhood-onset major depression: preliminary findings. Archives of General Psychiatry 2002;59:173-179.

[17] van der Veen FM, Evers EA, Deutz NE, Schmitt JA. Effects of acute tryptophan depletion on mood and facial emotion perception related brain activation and performance in healthy women with and without a family history of depression. Neuropsychopharmacology 2007;32:216-224.

[18] Gotlib IH, Hamilton JP, Cooney RE, Singh MK, Henry ML, Joormann J. Neural processing of reward and loss in girls at risk for major depression. Archives of General Psychiatry 2010;67:380-387.

[19] Vythilingam M, Heim C, Newport J, Miller AH, Anderson E, Bronen R, et al. Childhood trauma associated with smaller hippocampal volume in women with major depression. American Journal of Psychiatry 2002;159:2072-2080.

[20] van Harmelen AL, van Tol MJ, van der Wee NJ, Veltman DJ, Aleman A, Spinhoven P, et al. Reduced medial prefrontal cortex volume in adults reporting childhood emotional maltreatment. Biological Psychiatry 2010;68:832-838.

[21] Keyes CL, Myers JM, Kendler KS. The structure of the genetic and environmental influences on mental wellbeing. American Journal of Public Health 2010;100:2379-2384.

[22] Papadakis AA, Prince RP, Jones NP, Strauman TJ. Self-regulation, rumination, and vulnerability to depression in adolescent girls. Development and Psychopathology 2006;18:815-829.

[23] Pietrzak RH, Johnson DC, Goldstein MB, Malley JC, Rivers AJ, Morgan CA, et al. Psychosocial buffers of traumatic stress, depressive symptoms, and psychosocial difficulties in veterans of Operations Enduring Freedom and Iraqi Freedom: the role of resilience, unit support, and postdeployment social support. Journal of Special Operations Medicine 2009;9:74-78.

[24] Spijker J, de Graaf R, Oldehinkel AJ, Nolen WA, Ormel J. Are the vulnerability effects of personality and psychosocial functioning on depression accounted for by subthreshold symptoms? Depression and Anxiety 2007;24:472-478.

[25] Cremers HR, Demenescu LR, Aleman A, Renken R, van Tol MJ, van der Wee NJ, et al. Neuroticism modulates amygdala-prefrontal connectivity in response to negative emotional facial expressions. Neuroimage 2010;49:963-970.

[26] Alexopoulos GS, Kiosses DN, Choi SJ, Murphy CF, Lim KO. Frontal white matter microstructure and treatment response of late-life depression: a preliminary study. American Journal of Psychiatry 2002;159:1929-1932.

[27] Etkin A, Pittenger C, Polan HJ, Kandel ER. Toward a neurobiology of psychotherapy: basic science and clinical applications. Journal of Neuropsychiatry and Clinical Neurosciences 2005;17:145-158.

[28] Mayberg HS, Brannan SK, Mahurin RK, Jerabek PA, Brickman JS, Tekell JL, et al. Cingulate function in depression: a potential predictor of treatment response. Neuroreport 1997;8:1057-1061.

[29] McGuire PK, Bench CJ, Frith CD, Marks IM, Frackowiak RS, Dolan RJ. Functional anatomy of obsessive-compulsive phenomena. British Journal of Psychiatry 1994;164:459-468.

[30] Rauch SL, Jenike MA, Alpert NM, Baer L, Breiter HC, Savage CR, et al. Regional cerebral blood flow measured during symptom provocation in obsessive-compulsive disorder using oxygen 15-labeled carbon dioxide and positron emission tomography. Archives of General Psychiatry 1994;51:62-70.

[31] Gong Q, Wu Q, Scarpazza C, Lui S, Jia Z, Marquand A, et al. Prognostic prediction of therapeutic response in depression using high-field MR imaging. Neuroimage 2010;10:27-36.

[32] Keedwell PA, Drapier D, Surguladze S, Giampietro V, Brammer M, Phillips M. Subgenual cingulate and visual cortex responses to sad faces predict clinical outcome during antidepressant treatment for depression. Journal of Affect Disorders 2010;120:120-125.

[33] Farb NA, Anderson AK, Mayberg H, Bean J, McKeon D, Segal ZV. Minding one's emotions: mindfulness training alters the neural expression of sadness. Emotion 2010;10:25-33.

[34] Illes J, Lombera S, Rosenberg J, Arnow B. In the mind's eye: provider and patient attitudes on functional brain imaging. Journal of Psychiatric Research 2008;43:107-114.

[35] Hahn T, Marquand AF, Ehlis AC, Dresler T, Kittel-Schneider S, Jarczok TA, et al. Integrating Neurobiological Markers of Depression. Archives of General Psychiatry 2010:33;15-23.

[36] Sheline YI, Gado MH, Price JL. Amygdala core nuclei volumes are decreased in recurrent major depression. Neuroreport 1998;9:2023-2028.

[37] Mervaala E, Föhr J, Könönen M, Valkonen-Korhonen M, Vainio P, Partanen K, et al. Quantitative MRI of the hippocampus and amygdala in severe depression. Psychological Methods 2000;30:117-125.

[38] Shah PJ, Ebmeier KP, Glabus MF, Goodwin GM. Cortical grey matter reductions associated with treatmentresistant chronic unipolar depression. Controlled magnetic resonance imaging study. British Journal of Psychiatry 1998;172:527-532.

[39] Bremner JD, Narayan M, Anderson ER, Staib LH, Miller HL, Charney DS. Hippocampal volume reduction in major depression. American Journal of Psychiatry 2000;157:115-118.

[40] Greenwald BS, Kramer-Ginsberg E, Bogerts B, Ashtari M, Aupperle P, Wu H, et al. Qualitative magnetic resonance imaging findings in geriatric depression. Possible link between later-onset depression and Alzheimer's disease? Psychological Methods 1997;27:421-431.

[41] Keedwell PA, Andrew C, Williams SC, Brammer MJ, Phillips ML. The neural correlates of anhedonia in major depressive disorder. Biological Psychiatry 2005;58:843-853.

[42] Matthews SC, Strigo IA, Simmons AN, Yang TT, Paulus MP. Decreased functional coupling of the amygdala and supragenual cingulate is related to increased depression in unmedicated individuals with current major depressive disorder. Journal of Affective Disorders 2008;111:13-20.

[43] Sabatinelli D, Fortune EE, Li Q, Siddiqui A, Krafft C, Oliver WT, et al. Emotional perception: meta-analyses of face and natural scene processing. Neuroimage 2011;54:2524-2533.

[44] Ohman A. The role of the amygdala in human fear: automatic detection of threat. Psychoneuroendocrinology 2005;30:953-958.

[45] Winston JS, O'Doherty J, Dolan RJ. Common and distinct neural responses during direct and incidental processing of multiple facial emotions. Neuroimage 2003;20:84-97.

[46] Gorno-Tempini ML, Pradelli S, Serafini M, Pagnoni G, Baraldi P, Porro C, et al. Explicit and incidental facial expression processing: an fMRI study. Neuroimage 2001;14:465-473.

[47] Ochsner KN, Gross JJ. The cognitive control of emotion. Trends in Cognitive Sciences 2005;9:242-249.

[48] Phillips ML, Ladouceur CD, Drevets WC. A neural model of voluntary and automatic emotion regulation: implications for understanding the pathophysiology and neurodevelopment of bipolar disorder. Molecular Psychiatry 2008;13:829, 833-857.

[49] Carmichael ST, Price JL. Limbic connections of the orbital and medial prefrontal cortex in macaque monkeys. Journal of Comparative Neurology 1995;363:615-641.

[50] Coffey CE, Wilkinson WE, Weiner RD, Parashos IA, Djang WT, Webb MC, et al. Quantitative cerebral anatomy in depression. A controlled magnetic resonance imaging study. Archives of General Psychiatry 1993;50:7-16.

[51] Drevets WC, Price JL, Simpson JR Jr, Todd RD, Reich T, Vannier M, et al. Subgenual prefrontal cortex abnormalities in mood disorders. Nature 1997;386:824-827.

[52] Rajkowska G, Miguel-Hidalgo JJ, Wei J, Dilley G, Pittman SD, Meltzer HY, et al. Morphometric evidence for neuronal and glial prefrontal cell pathology in major depression. Biological Psychiatry 1999;45:1085-1098.

[53] Elliott R, Baker SC, Rogers RD, O'Leary DA, Paykel ES, Frith CD, et al. Prefrontal dysfunction in depressed patients performing a complex planning task: a study using positron emission tomography. Psychological Medicine 1997;27:931-942.

[54] Siegle GJ, Thompson W, Carter CS, Steinhauer SR, Thase ME. Increased amygdala and decreased dorsolateral prefrontal BOLD responses in unipolar depression: related and independent features. Biological Psychiatry 2007;61:198-209.

[55] Davidson RJ, Pizzagalli D, Nitschke JB, Putnam K. Depression: perspectives from affective neuroscience. Annual Review Psychology 2002;53:545-574.

[56] Bench CJ, Friston KJ, Brown RG, Frackowiak RS, Dolan RJ. Regional cerebral blood flow in depression measured by positron emission tomography: the relationship with clinical dimensions. Psychological Methods 1993;23:579-590.

[57] Erk S, Mikschl A, Stier S, Ciaramidaro A, Gapp V, Weber B, et al. Acute and sustained effects of cognitive emotion regulation in major depression. Journal of Neuroscience 2010;30:15726-15734.

[58] Vasic N, Wolf RC, Walter H. Executive functions in pa-tients with depression. The role of prefrontal activation. Nervenarzt 2007;78:628-636.

[59] Amini F, Lewis T, Lannon R, Louie A, Baumbacher G, McGuinness T, et al. Affect, attachment, memory: contributions toward psychobiologic integration. Psychiatry 1996;59:213-239.

[60] Aggleton JP. The Amygdala: neurobiological aspect of emotion, memory, and mental dysfunction. New work: Wiley-Liss;2000. p1- 66.

[61] Bouton ME. A learning theory perspective on lapse, relapse, and the maintenance of behavior change. Health Psychology 2000;19:57-63.

[62] Beutel ME, Stern E, Silbersweig DA. The emerging dialogue between psychoanalysis and neuroscience: neuroimaging perspectives. Journal of the American Psychoanalytic Association 2003;51:773–801.

[63] Anderson MC, Ochsner KN, Kuhl B, Cooper J, Robertson E, Gabrieli SW, Glover GH, Gabrieli JD. Neural systems underlying the suppression of unwanted memories. Science 2004;303:232–235.

[64] Bunge SA, Ochsner KN, Desmond JE, Glover GH, Gabrieli JD. Prefrontal regions involved in keeping information out of mind. Brain 2001;124: 2074–2086.

[65] Andreasen NC, O'Leary DS, Cizadlo T, Arndt S, Rezai K, Watkins GL, Ponto LL, Hichwa RD. Remembering the past: two facets of episodic memory explored with positron emission tomography. American Journal of Psychiatry 1995;152: 1576–1585.

[66] Ochsner KN, Bunge SA, Gross JJ, Gabrieli JD. Rethinking feelings : an FMRI study of the cognitive regulation of emotion. Journal of Cognitive Neuroscience 2002;14:1215–1229.

[67] Ochsner KN, Feldman, Barrett L. A multi-process perspective on the neuroscience of emotion. In Emotion: Current Issues and Future Directions (ed. R. J. Davidson, H. Goldsmith and K. R. Scherer). New York; Oxford University Press: 2001.

[68] Milad MR, Rauch SL, Pittman RK, Quirk GJ. Fear extinction in rats : implications for human brain imaging and anxiety disorders. Biological Psychiatry 2003;8:572-580..

[69] Maren S, Quirk GJ. Neuronal signaling of fear memory. Nature Reviews Neuroscience 2004;5:844–852.

[70] Buchel C, Dolan RJ, Armony JL, Friston KJ. Amygdala-hippocampal involvement in human averse trace conditioning revealed through event-related functional magnetic resonance imaging. Journal of Neuroscience 1999;19:10869–10876.

[71] Gottfried JA, Dolan RJ. Human orbitofrontal cortex mediates extinction learning while accessing conditioned representations of value. Nature Neuroscience 2004;7:1144–1152.

[72] Ablon JS, Jones EE. Validity of controlled trials of psychotherapy: findings from the NIMH Treatment of Depression Collaborative Research Program. American Journal of Psychiatry 2002;159:775–783.

[73] Giove F, Mangia S, Bianciardi M, Garreffa G, DiSalle F, Morrone R, Maraviglia B. The physiology and metabolism of neuronal activation : in vivo studies by NMR and other methods. Magnetic Resonance Imaging 2003;21:1283–1293.

[74] Dougherty DD, Rauch SL, Rosenbaum JF. Essentials of neuroimaging for clinical practice. Washington: American Psychiatric Association; 2004.

[75] DeRubeis RJ, Siegle GJ, Hollon SD. Cognitive therapy versus medication for depression: treatment outcomes and neural mechanisms. Nature Reviews Neuroscience 2008;9:788–796.

[76] Frewen PA, Dozois DJ, Lanius RA. Neuroimaging studies of psychological interventions for mood and anxiety disorders: empirical and methodological review. Clinical Psychology Review 2008;28:228– 246.

[77] Beck AT. The evolution of the cognitive model of depression and its neurobiological correlates. American Journal of Psychiatr 2008;165:969–977.

[78] Kirsch I, Deacon BJ, Huedo-Medina TB, Scoboria A, Moore TJ, Johnson BT. Initial severity and antidepressant benefits: a meta-analysis of data submitted to the Food and Drug Administration. PLoS Medicine 2008;5:45.

[79] Drevets WC. Prefrontal cortical-amygdalar metabolism in major depression. Annals of the New York Academy Sciences 1999;877:614–637.

[80] Sheline YI, Barch DM, Donnelly JM, Ollinger JM, Snyder AZ, Mintun MA. Increased amygdala response to masked emotional faces in depressed subjects resolves with antidepressant treatment: an fMRI study. Biological Psychiatry 2001;50:651–658.

[81] Abler B, Erk S, Herwig U, Walter H. Anticipation of aversive stimuli activates extended amygdala in unipolar depression. Journal of Psychiatric Research 2007;41:511–522.

[82] Baxter LR Jr, Schwartz JM, Phelps ME, Mazziotta JC, Guze BH, Selin CE, Gerner RH, Sumida RM. Reduction of prefrontal cortex glucose metabolism common to three types of depression. Archives of General Psychiatry 1989;46:243–250.

[83] Mayberg HS, Liotti M, Brannan SK, McGinnis S, Mahurin RK, Jerabek PA, Silva JA, Tekell JL, Martin CC, Lancaster JL, Fox PT. Reciprocal limbic-cortical function and negative mood: converging PET findings in depression and normal sadness. American Journal of Psychiatry 1999;156:675–682.

[84] Mayberg HS. Modulating dysfunctional limbic-cortical circuits in depression: towards development of brain-based algorithms for diagnosis and optimised treatment. British Medical Bulletin 2003;65:193–207.

[85] Anand A, Li Y, Wang Y, Gardner K, Lowe MJ. Reciprocal effects of antidepressant treatment on activity and connectivity of the mood regulating circuit: an FMRI study. Journal of Neuropsychiatry and clinical neurosciences 2007;19:274–282.

[86] Mayberg HS, Silva JA, Brannan SK, Tekell JL, Mahurin RK, McGinnis S, Jerabek PA. The functional neuroanatomy of the placebo effect. American Journal of Psychiatry 2002;159:728–737.

[87] Goldapple K, Segal Z, Garson C, Lau M, Bieling P, Kennedy S, Mayberg H. Modulation of cortical-limbic pathways in major depression. Archives of General Psychiatry 2004;61:34–41.

[88] Paquette V, Levesque J, Mensour B, Leroux JM, Beaudoin G, Bourgouin P, Beauregard M. 'Change the mind and you change the brain': effects of cognitive-behavioral therapy on the neural correlates of spider phobia. Neuroimage 2003;18: 401–409.

[89] Ochsner KN, Ray RD, Cooper JC, Robertson ER, Chopra S, Gabrieli JD, Gross JJ. For better or worse: neural systems supporting the cognitive down- and up-regulation of negative emotion. Neuroimage 2004;23:483–499.

[90] Friston K. Functional and effective connectivity in neuroimaging: a synthesis. Human Brain Mapping 1994;2:58–78.

[91] Horwitz B, Tagamets MA, McIntosh AR. Neural modeling, functional brain imaging, and cognition. Trends in Cognitive Sciences 1999;3:91–98.

[92] Seminowicz DA, Mayberg HS, McIntosh AR, Goldapple K, Kennedy S, Segal Z, Rafi-Tari S. Limbic-frontal circuitry in major depression: a path modeling metanalysis. Neuroimage 2004;22:409–418.

[93] Kennedy SH, Konarski JZ, Segal Z, Lau MA, Bieling PJ, McIntyre RS, Mayberg HS. Differences in brain glucose metabolism between responders to CBT and venlafaxine in a 16-week randomized controlled trial. American Journal of Psychiatry 2007;164:778–788.

[94] Fu CH, Williams SC, Cleare AJ, Brammer MJ, Walsh ND, Kim J, Andrew CM, Pich EM, Williams PM, Reed LJ, Mitterschiffthaler MT, Suckling J, Bullmore ET. Attenua-

tion of the neural response to sad faces in major depression by antidepressant treatment: a prospective, event-related functional magnetic resonance imaging study. Archives of General Psychiatry 2004; 61:877–889.

[95] Fu CH, Williams SC, Cleare AJ, Scott J, Mitterschiffthaler MT, Walsh ND, Donaldson C, Suckling J, Andrew C, Steiner H, Murray RM. Neural responses to sad facial expressions in major depression following cognitive behavioral therapy. Biological Psychiatry 2008;64:505–512.

[96] Martin SD, Martin RN, Rai SS, Richardson MA, Royall R, Eng IE. Brain blood flow changes in depressed patients treated with interpersonal psychotherapy or venlafaxine hydrochloride: preliminary findings. Archives of General Psychiatry 2001; 58: 641–648.

[97] Brody AL, Saxena S, Stoessel P, Gillies LA, Fairbanks LA, Alborzian S, Phelps ME, Huang SC, Wu HM, Ho ML, Ho MK, Au SC, Maidment K, Lewis R. Regional brain metabolic changes in patients with major depression treated with either paroxetine or interpersonal therapy: preliminary findings. Archives of General Psychiatry 2001;58, 631–640.

[98] Brody AL, Saxena S, Silverman DH, Alborzian S, Fairbanks LA, Phelps ME, Huang SC, Wu HM, Maidment K, Baxter LR. Brain metabolic changes in major depressive disorder from pre- to post-treatment with paroxetine. Psychiatry Research 1999;13:27–139.

[99] Brody AL, Saxena S, Mandelkern MA, Fairbanks LA, Ho ML, Baxter LR. Brain metabolic changes associated with symptom factor improvement in major depressive disorder. Biological Psychiatry 2001;50:171–178.

[100] Buchheim A, Viviani R, Kessler H, Kächele H, Cierpka M, Roth G, George C, Kernberg OF, Bruns G, Taubner S. Changes in prefrontallimbic function in major depression after 15 months of long-term psychotherapy. PLoS One 2012; 7:e33745.

[101] Zantvoord JB, Diehle J, Lindauer RJ. Using neurobiological measures to predict and assess treatment outcome of psychotherapy in posttraumatic stress disorder: systematic review. Psychotherapy and Psychosomatics 2013; 82: 142–151.

[102] Konarski JZ, Kennedy SH, Segal ZV, Lau MA, Bieling PJ, McIntyre RS, Mayberg HS. Predictors of nonresponse to cognitive behavioural therapy or venlafaxine using glucose metabolism in major depressive disorder. Journal of Psychiatry and Neuroscience 2009; 34: 175–180.

[103] Siegle GJ, Thompson WK, Collier A, Berman SR, Feldmiller J, Thase ME, Friedman ES. Toward clinically useful neuroimaging in depression treatment: prognostic utility of subgenual cingulate activity for determining depression outcome in cognitive therapy across studies, scanners, and patient characteristics. Archives of General Psychiatry 2012; 69: 913–924.

[104] Hirvonen J, Hietala J, Kajander J, Markkula J, Rasi-Hakala H, Salminen J, Nagren K, Aalto S, Karlsson H. Effects of antidepressant drug treatment and psychotherapy on striatal and thalamic dopamine D2/3 receptors in major depressive disorder studied with [11c]raclopride PET. Journal of Psychopharmacology 2010;5(10);1329–1336.

[105] Karlsson H, Hirvonen J, Kajander J, Markkula J, Rasi-Hakala H, Salminen JK, Nagren K, Aalto S, Hietala J. Research letter: psychotherapy increases brain serotonin 5-HT1A receptors in patients with major depressive disorder. Psychological Medicine 2010;40: 523–528.

[106] Bhagwagar Z, Rabiner EA, Sargent PA, Grasby PM, Cowen PJ, Persistent reduction in brain serotonin1A receptor binding in recovered depressed men measured by positron emission tomography with [11C]WAY-100635. Molecular Psychiatry 2004; 9:386–392.

[107] Drevets WC, Frank E, Price JC, Kupfer DJ, Holt D, Greer PJ, Huang Y, Gautier C, Mathis C. PET imaging of serotonin 1A receptor binding in depression. Biological Psychiatry 1999;46:1375–1387.

[108] Sargent PA, Kjaer KH, Bench CJ, Rabiner EA, Messa C, Meyer J, Gunn RN, Grasby PM., Cowen PJ. Brain serotonin1A receptor binding measured by positron emission tomography with [11C]WAY-100635: effects of depression and antidepressant treatment. Archives of General Psychiatry 2000;57: 174–180.

[109] Laruelle M, Slifstein M, Huang Y. Positron emission tomography: imaging and quantification of neurotransporter availability. Methods 2002;27: 287–299.

[110] Dichter GS, Felder JN, Smoski MJ. Affective context interferes with cognitive control in unipolar depression: an fMRI investigation. Journal of Affective Disorders 2009;114, 131–142.

[111] Hopko DR, Lejuez CW, Ruggiero KJ, Eifert GH. Contemporary behavioral activation treatments for depression: procedures, principles, and progress. Clinical Psychology Review 2003;23, 699–717.

[112] Dichter GS, Felder JN, Smoski MJ. The effects of Brief Behavioral Activation Therapy for Depression on cognitive control in affective contexts: an fMRI investigation. Journal of Affective Disorders 2010;126:236–244.

Chapter 5

The Neural Mechanism of Negative Cognitive Bias in Major Depression — Theoretical and Empirical Issues

Zhengzhi Feng, Xiaoxia Wang, Keyu Liu, Xiao Liu,
Lifei Wang, Xiao Chen and Qin Dai

Additional information is available at the end of the chapter

http://dx.doi.org/10.5772/59374

1. Introduction

Depression is one of the most prevalent mood disorders worldwide which threatens human mental health and well-being. Major depressive disorder is characterized by excessive negative mood and reduced experience of pleasure (anhedonia). The etiology of depression is much too complicated to explain when a wide spectrum of cognitive, affective, neurobiological symptoms taken into consideration.

According to Beck's influential cognitive theory, depression is characterized by presence of negative schemas, defined as mental representations of past experiences, containing dysfunctional attitudes about the self. These underlying schemas have an important influence on the way information is processed, guiding one's attention, memory and interpretation for personally relevant negative experiences [1]. Recent studies have proved that the negative schema can produce cognitive deficits such as negative attentional bias, over general autobiographical memory, cognitive control deficits. These cognitive deficits make people with depression inclined to choose negative material consistent with their negative schema, which cause persistent and recurrent depressive episodes. Other researchers used a variant of behavior paradigms, and neuroscience techniques such as Event-Related Potential (ERP) and Functional Magnetic Resonance Imaging (fMRI) to delineate the negative attentional bias, emotion dysregulation, over-general autobiographical memory, cognitive control deficits underpinning depression. However, several important questions remain. How do people with depression develop cognitive deficits? Can cognitive deficits predict the onset and process of depression? Can we fix the cognitive deficits of depressed individuals? What are the neural substrates underlying these cognitive deficits? If we can answer these questions, depression will be understood much better than what we do now. Specifically, to answer these questions, we have to keep an eye on several domains below.

Attention is the first step of cognitive process, but what role does it play in the negative cognitive bias? There is a large and growing literature trying to determine the cognitive mechanism that may underlie such attentional inflexibility in depressed individuals. Recent studies have found that people with depression are inclined to avoid from positive stimuli while difficult to disengage attention from negative stimuli in dot probe detection task [2]]. Similarly, people with depression cannot rule out the impact of emotion on the performance, their reaction time (RT) is prolonged when they are faced with sad faces [3]. Besides, studies using ERP have shown that when people with depression are presented with sad faces, their P3 will be delayed and stronger which means that dysphoric people pay more attention to negative stimulus [4], reflecting more elaborate processing of negative stimuli. These results indicate that people with depression have negative attentional bias which make people with depressive symptoms tend to focus on the negative stimulus, which consequently lead to sustained and exacerbated symptoms of depression.

Emotion appraisal and regulation are two essential systems underpinning the process of emotional cognition. It is widely assumed that the initial appraisals of emotion events are the starting point for iterative cycles of appraising and reappraising that extend beyond the events themselves. Thus the initial appraisals must play an essential role in the negative bias of depression. Emotional appraisal is described as a multidimensional neural processing including sensory inputs and affective experience. To be noted, Emotion Context Insensitivity Hypothesis proposes that major depression is characterized by flattened emotional reactivity to both positive and negative valenced stimuli, with the reduction greater for positive stimuli [5]. Thus the positive and negative emotion system might be two relatively independent and interacting processes that need further exploration.

Autobiographical memory is a memory system consisting of episodes recollected from an individual's life, consisting of both episodic and semantic memory. Over general autobiographical memory is a key symptom of depression which is sustained over episodes of depression. Longitudinal studies focused on currently depressed patients and high-risk population found that the more over general autobiographical memory is, the longer depression is sustained. As a result, some researchers proposed that over general autobiographical memory play an essential role in the etiology of depression. Large amounts of researches have been done to delineate healthy people's autobiographical memory. Recent studied using behavioral methods have shown that the inhibition of other unrelated memory play an important role in the construction of autobiographical memory. Besides, studies using fMRI have discovered related brain circuits underpinning autobiographical memory [6]. Although much work has been done, some questions still remain: how does the depressive brain activate when it try to recall an autobiographical memory? You can read the attempts to find the answers in the following part of this chapter.

Disturbed interface between cognitive control and emotion processing. Recent research has suggested that impairments in memory and attention are related to cognitive control deficits. Consequently, persistent sad mood may be maintained by negative attentional bias. Meanwhile, anhedonia or flattened emotion pattern may be linked with abnormal processing of positive stimuli. Thus the cause of depression may be linked with altered emotional processing or impaired cognitive control. It remained largely unknown whether emotional processing and cognitive control affect the negative cognitive bias independently or reciprocally. Researches in this area have suggested that the negative cognitive bias reflect the over activated bottom-up system, which is related to over activated amygdala, fusiform gyrus and enlarged early P300 in ERP studies. Meanwhile, other researchers indicated that top-down cognitive control deficit can explain depression's negative cognitive bias. This deficit is associated with the hypo-activated dorsolateral prefrontal cortex (DLPFC), dorsal anterior cingulated cortex (dACC), rostral anterior cingulated cortex (rACC) and the smaller N450 and N200 in ERP studies [7]. Most researches explain negative cognitive bias either from down-top prospect (amygdala, fusiform gyrus) or from top-down perspective (e.g. DLPFC, dACC). But results of some studies indicate that these two system work together to form the negative cognitive bias. When negative information is processed by people with depression, bottom-up system may be over activated (e.g. over activated amygdala), while the weakened inhibition function of the top-down system may exacerbate this process and form a maladaptive circle which may lead to a depressive episode finally.

Although much effort has been made to uncover the mystery of depression, many questions about the depression's neural basis still remain. Firstly, we cannot find a valuable predictor to the depressive episode's onset yet. Secondly, the association between negative cognitive bias and the specific brain deficits (neural circuit) is not yet quite clear. Not until recently, we just began to understand the neural process underlying the cognitive training and rTMS directed to depression. Though these questions seem difficult to answer, they may be the key for us to conquer depression in the future.

2. The neurocognitive mechanisms behind attentional bias in depression

According to Beck's theory, individuals' existing memory representations, or schemas, made their attention directed to the information that is congruent with their own schemas, and initiated a vicious cycle of negative automatic thoughts, processing biases, and depressed mood [8]. Moreover, differential attentional bias to negative information has been considered a cognitive vulnerability factor for the development of depression [9]. For this reason, attentional bias have drawn researchers' attention and been supported by various studies.

2.1. Theories on attentional bias in depression and limitations

Attentional bias could be understood from the perspectives of attention process, narrow focus of attention, cognitive load and arousal level [10].

Attention processes include orienting, maintenance and disengagement, and it is still controversial to say which progress leads to attentional bias. For one explanation, attention may be attracted to or directed away from the position of negative stimuli in the orienting progress [11]. For another, it is the attention disengagement that emotional stimuli affects, making maintained attention to the negative stimuli, or being relatively easier distracted from positive stimuli [9]. It is significant to discuss from a dynamic point of view, but it has its own limitation for further study on specific mechanism.

Theories on narrow focus of attention refers to that depressed patients may selectively attend to depression-related information and process mood-congruent material easier, while ignoring other information, which represents attention bias. This interpretation represents neural network specifity while ignores the control ability of neural pathways for information processing [12].

When it comes to **cognitive capacity theory**, some postulate that individuals can only perform limited mental activity due to fixed amount of cognitive resource. When negative stimuli appear with other stimuli, depressed individuals' attentional resource can easily be occupied with the negative one, leaving little cognitive resources to other tasks and making them uncompleted. Other experts then believe depressed ones are more engaged in their *internal mental activity*, while they are less sensitive to external stimuli. Sometimes they only respond to intense stimuli. This can be well explained by *parallel distributed processing (PDP) model* [13]. Research showed that stimulus which was either too weak or not associated with depression was not reach the arousal level of subjects and therefore no attentional bias was observed [14].

Here come limitations with these existing explanations. Firstly, it is unclear that which mechanism, facilitation or inhibition, leads to attentional bias. Selective attention means processing information selectively which lead to activation for relevant information and inhibition for irrelevant or interferential information. Or maybe the cooperation of two mechanisms makes selective attention. Secondly, what is the relationship between attentional bias and depression? It is attentional bias that cause depression or just in reverse. Or maybe attentional bias is just one of cognitive features of depression.

2.2. Researches on attentional bias

To further understand these questions, we need firstly get a picture of research paradigms in this field as well as the features of attentional bias in depressed patients.

2.2.1. Research paradigms for attentional bias

Stroop task, or emotionally modified stroop task, is the principal paradigm for attentional bias. By measuring subjects' reaction time performance in color-naming task or emotional words and its accuracy, interference inhibition is appraised indirectly.

Dot-probe detection task help demonstrate orienting or maintenance of attention. If subject had a shorter reaction time for target stimulus presented following a negative stimulus, he may have the attentional bias for negative stimuli.

Cue-target paradigm can observe orienting and disengagement of attention (Fig. 1). With emotional stimuli being valid and invalid cues, if subjects have shorter reaction time to negative valid stimuli and longer reaction time to negative invalid stimuli, they may present attentional bias.

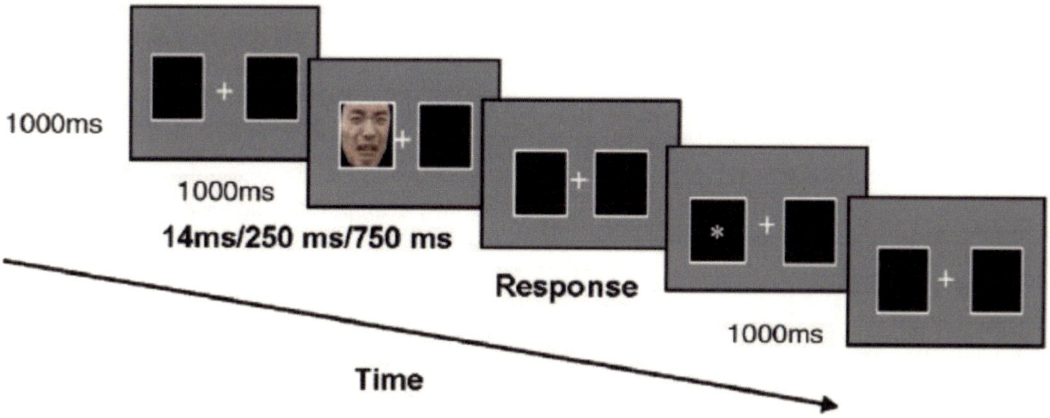

Figure 1. Cue-target task (one complete valid cue condition of sad faces) by Dai,2011, NATURE REVIEWS I NEURO-SCIENCE, 12, p.475. Copyright 2011 by Macmillan Publishers Limited

Garner paradigm is mainly used to investigate the maintenance of attention by studying the interaction of two variables, while the negative priming paradigms discuss attentional bias mainly from the angle of distracter inhibition. The latter task is designed to distinguish activation from inhibition which accounts of selective attention. The negative priming effect is defined as longer response latency when the distracter from a previous trial becomes the target on the present trial. And the distractor that activates in two trials may cause a delay response.

2.2.2. Features of attentional bias in depressed patients

Results of previous studies on the attentional bias of depression have been mixed, and the inconsistencies have been ascribed to differences among subjects, stimulus exposure duration, content and intensity. Therefore, features of attentional bias in depressed patients should be mentioned. Firstly, attentional bias to negative stimuli was found significant in researches with study subjects being clinical depressed patients. But when it comes to college students whose depression scores are relative high, biased processing of negative information is not significant [15]. Therefore, it is only when individuals get severe depression may significant attentional bias occur. Secondly, attentional bias occurs only when the stimuli last longer than the time threshold. In general, 1000ms is considered the best choice. For depressed patients, the internal attentional bias may lower their external stimuli arouse level and make it difficult to response to the stimuli from outside. Thus stimuli have been presented long enough for depressed individuals to aware and process. This has been shown in Bradley's and Gotlib's experiments. The former reported a mood-congruent bias on the dot-probe task under conditions of long stimuli exposures of 500 or 1000ms, but not under conditions of brief durations [16]. The latter had stimuli presented for 1000ms and observed attentional bias for negative faces in clinically diagnosed depressed patients [17]. Thirdly, the content of stimuli should be specific to depressed patients. Study with stimuli relevant to anxiety or self–esteem threat did not show attentional bias[18], but studies on sad faces or depression relevant words could induce negative attentional bias[11, 19]. The emotion-congruent information may strike a responsive chord in the heart of subjects who will show specific attentional bias for these negative stimuli. Fourthly, too weak stimulus intensity should be avoided. Emotional words, emotional faces, and emotional pictures are three types of stimuli that are widely used nowadays. Because emotional words always need deeper cognitive processing and emotional faces interpret more social information, these two types stimuli can induce more intense emotional experience. That is the reason why they are popular used. It seems that participants diagnosed with a current major depressive disorder (MDD) show attentional bias for long-exposed and very intense stimuli.

2.3. Cognitive neuroscience basis of attentional bias

In normal individuals, the anterior cingulate regulates the attention–emotion balance by signaling to the dorsolateral prefrontal cortex [20]. However, depressed individuals show less activation in these regions [21], which might relate to their inability to gain attentional control over emotional interference. Moreover, evidence showed that depressed subjects were more easily distracted by negative information, and spent significantly more time on processing these negative stimuli [22, 23]. Therefore, attentional control for emotional stimuli in depressed patients, especially deficient inhibition for attention to negative stimuli, needs further exploration.

Three kinds of attentional inhibitions have been explored: inhibition of return (IOR), distracter inhibition and interference inhibition.

IOR means a slower response to objects appearing at a formerly-attended location, which typically appears about 200–300 ms after stimulus offset. We investigated this phenomenon in depressed individuals with a cue target paradigm as emotional faces being cues in an event-related potential (ERP) experiment. Three groups of participants were recruited representing the health control (NC), the remitted depressive patients (RMD), and the major depressive disorder (MDD) group respectively. The MDD participants were found to have cue validity and deficient IOR for negative stimuli. The deficient inhibition of negative stimuli renders them unable to eliminate the interference of negative stimuli and causes the maintenance and development of depression [24].

To investigate distracter inhibition and facilitation for emotional faces in depressed individuals, we used a modified negative priming paradigm-Negative Affective Priming task (NAP) among three groups of subjects (NC, RMD and MDD). Though there were no significant differences among the three groups in the positive and negative priming effects of happy faces, differences between each pair of groups are significant for sad faces. Depressed individuals are found characterized by enhanced facilitation and deficient inhibition for negative materials, which is a stable cognitive vulnerability factor and is probably associated with occurrence of depression [25, 26].

Interference inhibition was investigated in current remitted depression using the emotional Stroop task and the event-related potential (ERP) technique. MDD participants had higher interference effects for negative words compared with the other two groups and those of positive stimuli. With regard to the ERP data, the MDD participants showed smaller N1 amplitude for negative words and a smaller P1 amplitude for positive words in bilateral hemispheres compared with the other groups. Both the MDD and RMD participants showed enhanced negativity (N450) over the parietal regions of the brain for negative words compared with NC groups [27].

In conclusion, negative attentional bias, an important contributor to the maintenance and development of depression, correlates with susceptibility to depression. The clinical psychologist could pay more attention on the training of intentional inhibition of negative stimuli during clinical intervention. These findings also provide a theoretical foundation for cognitive therapy, that is to say, patients can be cured by leading them to neglect or inhibit negative stimuli, or to pay more attention to positive ones. However, the correlation between attentional bias and depression remains unsolved. If attentional bias is causal, it is difficult to explain it in RMD participants.

3. Emotion appraisal and regulation disorder in depression

Emotion dysregulation is characteristic of MDD which occurred during an emotion generating episode. While emotion reactivity and regulation are two sides of a coin, emotion regulation is always explicitly or implicitly existent in emotion responses. Thus the need to change the initial negatively biased emotion responses is in particular essential for depressed patients.

3.1. Abnormal emotion reactivity

Loss of interest or pleasure is the core feature of depression, which may be manifested as lingering low mood and a reduced capacity to experience pleasure. Cognitive theories put maladaptive appraisal processes at the core of depression [28].According to Gross' s model of emotion, the initial appraisals of emotion events are frequently the starting point for iterative cycles of appraising and reappraising that extend beyond the events themselves[29]. Thus the need to change the initial negatively biased emotion responses is in particular essential for depressed patients.

Emotional appraisal has been presumed to span a hierarchy of neural processing, from sensory inputs to affective experience. Depression-related disruption of this hierarchical processing system may extending from sensory to frontal regions through insula [30].Researchers have proposes Emotion Context Insensitivity (ECI) hypothesis that major depression is characterized by flattened emotional reactivity to both positive and negative valenced stimuli, with the reduction greater for positive stimuli[31]. And ECI was supported in three main subsystems of emotion response: peripheral physiology, expressive behavior and self-reported arousal [32]. At resting state, amygdala-hippocampal/brainstem and amygdala-precuneus may be circuits that are important for modulation of physiologic responses to emotion which are impaired and contribute to both mood and vegetative symptoms [33]. During the completion of emotional intensity evaluation, we found that more excited perception for negative facial expressions is a stable cognitive vulnerability possibly associated with the occurrence or recurrence of depression [34]. This could partly explain a generalized emotional hypoactivity in major depression.

3.2. Disrupted emotion regulation

While emotion reactivity and regulation are two sides of a coin, emotion regulation is always explicitly or implicitly existent in emotion responses. Emerging evidence emphasizes the role of emotion regulation capacity in the neurological model of depression; depressed patients exhibit difficulties implementing adaptive emotion regulation strategies and more frequent use maladaptive strategies. However, the mechanisms underlying the difficulties in emotion regulation remain unclear.

A long tradition of cognitive theory focused on a negative bias which causes the vulnerability and maintenance of major depression. Along with this tradition, Joormann et al proposes that cognitive biases and deficits in cognitive control typical of depression influence emotion regulation in critical ways, therefore resulting in maintained negative affect and diminished levels of positive affect[35].The evidence was that difficulties with cognitive flexibility and control may impair performance on tasks that require processing of relevant emotional stimuli[36].Neurally, depression may impair the cognitive capacity of depressed patients by recruiting more brain resources than controls during cognitive control. Accordingly, neuroimaging studies indicated increased activity within subcortical and ventral prefrontal cortical regions to negative emotional stimuli and decreased activity within dorsal prefrontal cortical regions in MDD patients.

Seminowicz et al put forward a limbic – cortical dysregulation model which proposes that sadness and depressive illness are both associated with decreases in dorsal neocortical regions (sensory-cognitive compartment) and relative increases in ventral limbic and paralimbic areas (autonomic compartment), and illness remission occurs when there is appropriate modulation of dysfunctional limbic–cortical interactions[37].

Phillips and colleagues schematize the neural basis of emotion perception and regulation deficits for major depression with a dorsal-ventral interaction model. Volume reductions within the amygdala and other components of the ventral neural system, together with increased activity within these regions during illness, may result in flattened emotional response patterns, biased toward the biased perception of negative emotions for amygdala [38]. The negative affectivity bias of depression can be seen during completion of emotional intensity evaluation task; we found that more excited perception for negative facial expressions is a stable cognitive vulnerability possibly associated with the occurrence or recurrence of depression [39]. Notably, negative bias might also been seen in bipolar depression, suggesting different pathophysiologic processes for BD versus MDD depression. There was evidence that abnormally elevated left amygdala activity to mild sad and neutral faces might be a depression-specific marker in BD but not MDD, [40].Thus further exploration into emotion processing disruption and neural mechanisms for MDD is necessary.

Structural and functional impairments within regions of the dorsal system, associated with impairments in executive function and effortful regulation of emotional behavior, may perpetuate these phenomena, setting stage for lingering depressed mood and anhedonia. As an example, reappraisal ability of emotion stimuli is found to be impaired in depressed individuals. Patients seemed to have deficits in prefrontal-amygdala modulatory network during both up-regulation and down-regulation of emotion [41]. Other studies suggest that overregulation of the amygdala by the ventromedial prefrontal cortex/ orbitomedial prefrontal cortex may lead to diminished amygdala responsiveness to happy faces in unipolar depressed patients [40, 42]. This could partly explain a generalized emotional hypoactivity in major depression.

3.2.1. *Typical use of emotion regulation strategies*

In the last decade, researchers have focused on the habitual use of specific strategies, examining the relation between use of adaptive and maladaptive strategies and psychopathology [43]. This line of research provides evidence that depressed individuals show a more dysfunctional use of emotion regulation strategies than controls. Maladaptive strategies (rumination, suppression), compared to adaptive strategies (reappraisal, problem-solving), were more strongly associated with depressive symptoms [44].Across both male and female groups, higher reports of self-blame, rumination and/or catastrophizing as strategies were strongly related to higher depression scores, whereas higher reports of positive reappraisal were related to lower depression scores[45].Moreover, these deficits are not limited to the acute phase but are also a risk factor for the development of recurrent depressive episodes[46].MDD patients reported increased suppression of both negative and positive emotions. Suppression of negative and positive emotions was related to depressive symptoms. Results demonstrated that suppression produced short-term reductions in sadness, while for moderate and higher levels of anxiety about the experience of depressed mood suppression was no longer effective[47]. The explanation for why MDD patients suppress emotions might be the fear of strong emotion [48].

Past research has also paid attention to potential cultural and gender differences in emotion regulation [49]. For instance, the association between the use of reappraisal and depressive symptoms was significantly stronger in the Korean compared to the US sample. In contrast, the association between anger suppression and depressive symptoms was significantly stronger in the American compared to the Korean sample[50].Due to these factors leading to individual differences of ER, examining individual differences in the habitual use of emotion-regulation strategies may have remarkable potential to clarify emotion regulation (ER) models of MDDs[35].Recent findings suggest that individual differences in the use of emotion-regulation strategies play an important role in depression, and that deficits in cognitive control are associated with the use of maladaptive emotion-regulation strategies in this disorder.

3.2.2. *Experimental manipulated emotion regulation*

Some researchers study emotion regulation by instructing participants to engage in particular emotion-regulation strategy in response to an emotion eliciting stimulus and then observing the effects on participants 'subsequent emotions, cognitions, or physiological responding. Most of these studies have been conducted on the implementation of adaptive strategies, and this area clearly should be a focus of future research on ER in this disorder [43].Reappraisal is an adaptive emotion regulation strategy which involves the utilization of cognitive control to regulate semantic representations of affective stimuli. Findings are mixed when it comes to the question whether depressed people differ from their no depressed counterparts in their ability to reappraise. We inspect the role of self-perspective in reappraisal process of depressed patients in terms of goals (valence/arousal) and tactics (detachment/immersion). The results were that impaired modulatory effects of amygdala in depressed patients are compensated with strengthening cognitive control resources, with dissociable effects for different self-perspectives in reappraisal. This may help clarify the role of self-perspective underlying reappraisal in major depression [41]. We also found that depressed patients but not healthy controls enhanced their positive emotions while recruiting behavioral activation system (drive) underlying bilateral DLPFC. This may suggest a lack of control resources during generating positive emotion for depressed individuals, who are consequently more dependent on compensatory recruitment of control areas. Thus amplification of positive affect could be more difficult for depressed patients. These results fits nicely with the previously mentioned findings by Werner-Seidler et al.(2013) and Beblo et al. (2012) that depression is associated with fear of emotion and apprehension about experiencing intense emotion [51, 52].

3.2.3. Dispositional/ spontaneous emotion regulation

Depression-vulnerable individuals might also be expected to abnormal in their spontaneous use of particular forms of emotion regulation when depressed [53]. Habitual styles of emotion regulation may determine the automatic regulation pathway suggested by Phillips et al. (2008) [54]. Conversely, there was also evidence that unconscious emotion-regulation processes may interplay with conscious emotion-regulation processes to affect mental health [55].The paradigm of conscious emotion regulation was also problematic because participants may have underreported or inaccurately recalled their application of regulation strategies. Ehring et al proposed that spontaneous but not instructed emotion regulation play a more critical role in depression vulnerability [53].Thus it would be necessary to examine the relationships between emotion-regulation strategies and psychopathology simultaneously at the dispositional and state level.

3.3. Emotion regulation: associations with affective style

It was reported that affective style reflecting approach and inhibition might interact with ER strategies to influence depressed mood [56]. Some researchers have suggested that depression is associated with a reduced sensitivity to reward and an increased sensitivity to punishment [57]. Moreover, to reappraise stimuli as appetitive/ aversive is inherent in cognitive control process of emotion according to Gross's model of cognitive control of emotion (MCCE) [58]. Our results revealed that an altered motivational pattern of BAS hypoactivity and BIS hyperactivity [59]with overlapping prefrontal circuits with reappraisal, would differentially affected the neural responses underlying reappraisal. And BIS/BAS measures were associated with prefrontal/ amygdala activation during immersion but not detachment. Consequently, high self-focused cognitive and ruminative tendency (particularly negative) for emotion regulation of depressed patients was implied.

To examine the abnormal resting state functional connectivity in major depression, fALFF were compared between groups using REST and further correlated with BIS/BAS in SPSS18.0. In controls, fALFF of right cerebellum was negatively correlated with BIS, fALFF of left medial frontal gyrus were negatively correlated with BASD. In MDD, fALFF of left median frontal gyrus was positively correlated with BASF. Conclusions The abnormal fALFF pattern of MDD was identified in default mode network and mood regulation circuit, and correlated with BIS/ BAS index [60].

To conclude, there has been a long tradition of treatment for depression targeting emotion regulation malfunction. Different types of psychotherapy modulate aberrant emotion by engaging different but interacting pathways for emotion regulation. One promising therapy of Neurofeedback is attractive that it enables the patients themselves to voluntarily control their brain activity which increases their self-efficacy, which is an important therapeutic factor in many neuropsychiatric disorders [61]. During real-time fMRI-NF (Neurofeedback) training, participants receive feedback on their brain activity in real-time and are instructed to change this activation by instruction or imagery.

4. Over-generalization of autobiographical memory in depression

4.1. What is over-generalization of autobiographical memory?

Are you often absorbed in your happy reminiscences, such as your first date scene, or some bad moments like the day you lost your beloved dog? The specific reminiscences of past events you ever experienced or even more conceptual, self-related information are all called autobiographical memory [62].

You may have realized that autobiographical memory is just like a great album keeping our own moments important or not, some of which may influence us forever. Autobiographical memory is critical to the human experience, the influence of which mainly embodies in three aspects. First, it plays a key role in creating a sense of self and identity. Individuals could switch to their past times from the current time in the subject time dimension. Second, autobiographical memories serve as important guides for the future. As records of personal past experiences, such memories provide reminders of the lessons learned from them, thereby helping themselves or others to solve similar problems in the present or to plan for future action [63]. Third, it could help to reinforce social connection. For example, unfamiliar people can create trust and closeness with each other quickly by sharing similar life experiences.

Since Williams and Broadbent first described the overgeneral autobiographical memory (OGM) phenomenon using the Autobiographical Memory Test (AMT) to assess the specificity of autobiographical memory in the study of suicidal patients, a series of researches have replicated this phenomenon, indicating that OGM is a trait marker, as a predictor of the course of depression.

Autobiographical memory is defined as a memory of an event that occurred to him or her at a particular time and place and lasted less than one day (e.g., "my wedding ceremony"). In contrast, overgeneral autobiographical memories include categorical memories that refer to a class of generic events (e.g., "parties with my friends") and extended memories that refer to an event lasting more than one day (e.g., "when I was on vacation last month"). Individuals with depression or trauma-related anxiety disorder, such as posttraumatic stress disorder (PTSD) usually exhibit OGM that is more general memories, less specific memories.

4.2. Mechanisms underlying OGM

Conway and his colleagues proposed Self Memory System (SMS) model in respect of autobiographical memory(Conway & Pleydell-Pearce, 2000). It is supposed in this model that autobiographical memory representation is a continuous hierarchy(see Figure 2), ranging from 1) more broad, conceptual themes in the life story such as "work theme", "relationship theme", to 2) lifetime periods (e.g., "my studying time in college"),to 3) general events (e.g., "parties with friends"), to 4) event-specific knowledge (i.e., specific episodic memories, such as "the party with my friends on the day of our graduation") containing information about the sensory and perceptual aspects of unique events. A successful specific memory need to reach to the 4th level of the hierarchy (i.e. event-specific knowledge), usually via two processes: generative retrieval or direct retrieval. Generative retrieval is a top–down process down the SMS model hierarchy from the life story to event-specific knowledge bases one by one to specify the desired memory recollection. This course may take some efforts for individuals. In contrast, direct retrieval could be realized when event-specific knowledge is activated by cues in the environment, much easier than the former form of retrieval. More conceptual, intermediate representations (e.g., general events) are often activated during the early stages of generative retrieval that correspond to overgeneral memories.

Based on the Self Memory System model, the CaR-FA-X model was developed by Williams and colleagues to explain the mechanisms of OGM (see Figure 3). This model postulates that OGM results when the generative retrieval search process is aborted prematurely as a result of one or more of the three proposed mechanisms [64]. Three mechanisms delineated in the model may underlie OGM: capture and rumination, functional avoidance, and impaired executive control.

Capture and rumination are thought to occur when conceptual self-relevant information activates ruminative processes during retrieval, thereby "capturing" cognitive resources and disrupting the retrieval search. As we know, autobiographical memory is a kind of self-relevant long-term memory. Its retrieval is based on self-representation. It's common that depressed individuals have negative self-schema, which could easily activate emotional conceptual self-representation and ruminative processes. Therefore, more abstract, conceptual negative information corresponding with self-representation could be much more easily captured by the activated emotional self-representation. More cognitive resources might have been exhausted, resulting in insufficiency of cognitive resource available to access event-specific knowledge base.

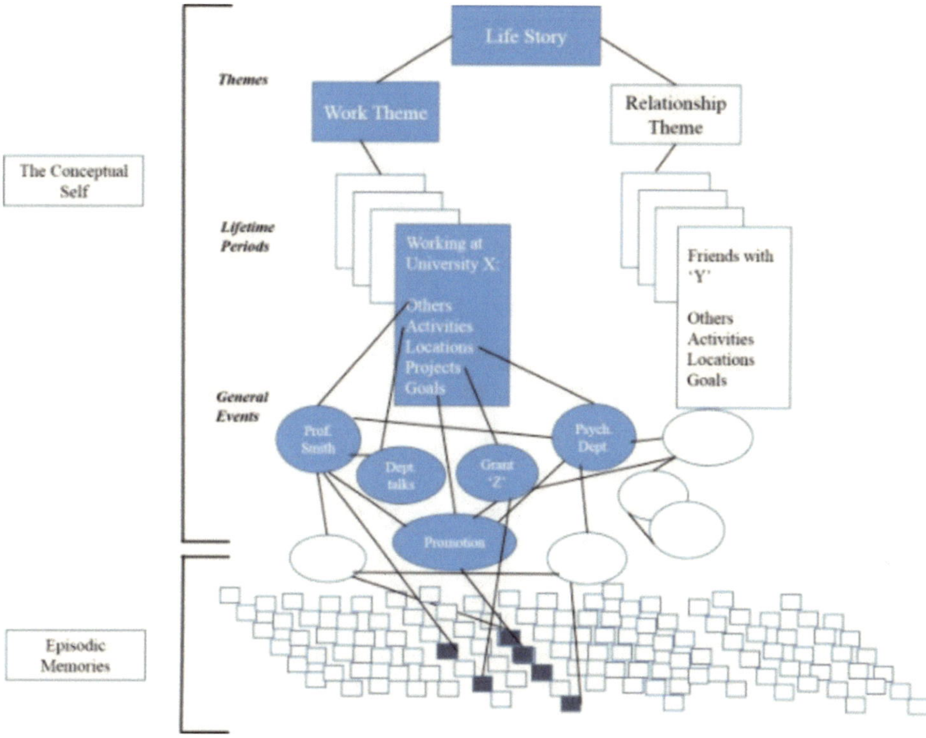

Figure 2. Hierarhcy of autobiographical memory representations From"Memory and the Self," by M. A. Conway, 2005, Journal ofMemory and Language, 53, p. 609. Copyright 2005 by Elsevier

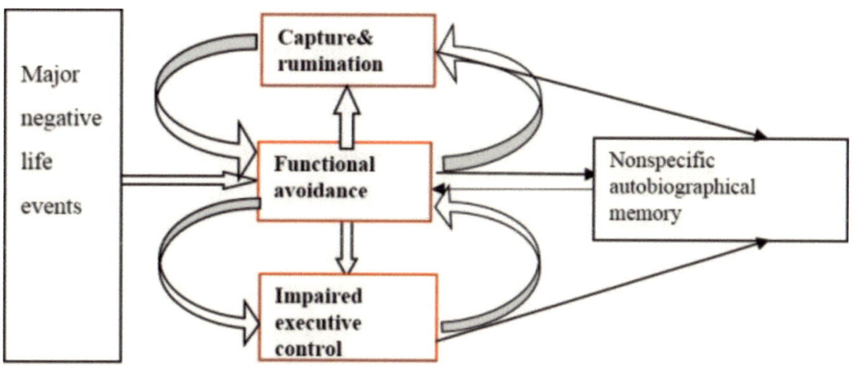

Figure 3. The CaR-FA-X model Three processes contributing to overgeneral memory-capture and rumination (CaR), functional avoidance (FA), and impaired executive capacity and control(X)-can each have effects on cognition and behavior,either independently or through their individual or combined effect on autobiographical memory. by Williams, 2007,Psychological Bulletin, 133, p.141. Copyright 2007 by the American Psychological Association

Functional avoidance refers to when the retrieval of specific memories is passively avoided as a means of affect regulation, and it is thought to occur in response to early trauma. It is hypothesized that specific details in memories could bring emotional distress about aversive experiences, the passive avoidance of which might reduce emotional distress to some extent. It sounds reasonable when the depressed people retrieve memories with negative cues. However, it cannot explain why the depressed people also exhibit OGM when retrieving with positive or neural cues.

Impaired executive control refers to when deficits in executive resources limit the ability to conduct a successful retrieval search. As mentioned, successful specific autobiographical memory retrieval involves strategic search processes that allow for current goals, the recovery of memory traces involving a rich sense of re-experience, and monitoring and other control processes [65]. Once executive control ability is impaired, it is hard for individuals to fulfill the AM task.

These mechanisms are hypothesized to contribute to OGM, alone or in interaction.

4.3. Instruments to measure OGM

As mentioned above, AMT designed by Williams has been regarded as a standard method to assess overgeneral autobiographical memory. On the AMT, individuals are usually visually or auditorially presented with cue words of different valences, and are asked to produce a specific memory and relevant details as many as possible related to the cue word within a given limit time (e.g., 30 s, 1min). Williams summarized that by using AMT, eleven studies had examined the specificity of memory in people suffering from major depressive disorder (MDD) [64]. He made a conclusion that almost all the studies successfully replicated overgeneral memory in depression.

However, Raes found the Sentence Completion for Events from the Past test (SCEPT), another method to measure OGM, more sensitive in non-clinical population, relative to AMT [66]. The SCEPT comprises several sentence stems (e.g. 'When I think of..."). Participants are instructed to complete the sentence stems with their past experiences. Further they found that the omission of the instruction to be specific was the probable reason to the enhanced sensitivity. Standard AMT usually had a much more detailed instruction, which may influence individuals' original autobiographical memory retrieval styles so as to decrease the sensitivity. So in our research we aimed to check the sensitivity of AMT without instruction, SCEPT, standard AMT and SCEPT with Specificity Instruction (SCEPT-SI) among Chinese healthy individuals or individuals with depressive mood. The results showed that under SCEPT without an explicit instruction, there was no significant difference between the mean proportions of general memory in the two groups (t=0.52, P>0.05), which was not associated with the BDI scores(r=0.96, P>0.05). Under AMT without explicit instructions, the mean proportion of general memory in the depressed mood group (DM) was significantly higher than that in the healthy control (HC) group (t=3.86, P<0.01), which was associated with depression scores(r=0.40, P<0.01). When the instruction was explicit, for the mean proportion of general memory on SCEPT-SI and the standard AMT, there were no significant difference between the two groups (P>0.05),and both of which were not associated with depression scores(respectively, r=-0.04, r=0.09, P>0.05). The conclusion in our research is that the sensitivity of AMT without explicit instruction was higher than that of standard AMT, which is in line with the previous finding Filip Rase et al [67]. However, that the sensitivity of SCEPT was higher than that of AMT without specific instruction was not found in the results. In this research, it was also suggested that SCEPT not sensitive for the non-clinical population to assess their over-general autobiographical memory.

4.4. Neural correlates and main processes of autobiographical memory

Just try to remember the last time you went to the bookstore. If it happened long ago, you should take more effort to search it in your brain. The constructive processing would be guided by the semantic information (e.g. Bookstores), and about your own life favor (e.g. I like bookstore with an awesome environment), even about inferential processes (e.g. I went with my roommate. He also loved reading). As the course went by, the memory that mostly fit into the requirement would be constructed. However, you should correct some incorrect information (e.g. the last time I went to the bookstore alone for receipt, rather than reading). At last, I could remember much detail about spatio-temporal, sensory and perception information and so on. Tulving supposed that autobiographical memory retrieval is a dynamic course [68]. When asked to begin a specific memory task, a participant should search relevant information in his/her autobiographical memory knowledge base in order to recover memory traces. At the same time, the participant should suppress irrelevant information, clarify uncertain information, and even correct the wrong. Once successfully constructed, the memory retrieval then moved to the elaboration phase to get as much detail as possible.

Functional neuroimaging studies have detected the neural correlates of autobiographical memory retrieval. Cabeza reviewed previous studies and summarized that the main processes of autobiographical memory retrieval may encompass constructive processes, monitoring processes, self-referential processes. And he also pointed out some characteristics of autobiographical memory (emotional processes, vividness, and remoteness). As to the neural correlates for each process: left lateral prefrontal cortex is associated with the constructive processes, and ventromedial PFC with monitoring processes, medial PFC associated with self-referential processes, amygdala associated with emotion processes, visual cortex associated with the vividness of AM, hippocampus associated with remoteness of AM[69]. Similarly, St. Jacques used independent component analysis (ICA) and found four separate neural networks supporting AM retrieval: medial prefrontal cortex (MPFC) network, associated with self-referential processes, medial temporal lobe (MTL) network, associated with memory, fronto-parietal network, associated with strategic search, and cingulo-operculum network, associated with goal maintenance [70]. Based on the results of normal healthy population, it is believed that OGM in depression probably has a close relationship with the failure in the main processes alone or in interaction. Lemogne proposed two modes of elevated MPFC activation in major depression, which may embody automatic aspects and strategic aspects of depressive self-focus respectively [71]. Dalgleish proved that depressed individuals exhibit OGM with impaired executive control ability in a series of behavioral experiment [72]. Whally discovered three regions of the prefrontal cortex: right inferior frontal gyrus, right middle frontal gyrus and left inferior associated with cognitive, emotional, and memory inhibition influencing specific autobiographical memories retrieval in depression using fMRI during an autobiographical memory task [73].

4.5. Application area

Research on OGM in depression is attracting increasing attention. Not only could it be a new and promising way to further discover the machanisms of depression, but it could provide a new target for treatment. As mentioned at the beginning, OGM is a trait marker as a predictor of the course of depression, which might be an important part in the early-warning and diagnosis of depression in future.

New cognitive traning has more recently been applied to a wide range of neuropsychiatric illnesses. One of which AM specificity training as a target-based tool to overcome OGM has already been paid attention and is expected to an effecitve method to cure depression.

Further research should deal with the neurocognitive mechanisms of OGM in depression. First, mechanisms of autobiographical memory and over-generalization of autobiographical memory call for further exploration. There is still no study to control one or more AM components directly to identify the role of each process in the course of AM retrieval. Moreover, if any molecular biological marker about AM was found, it would be helpful to targeted medication to depression. Second, more researches should be conducted on the effective cognitive traning methods like AM sepecificity training, which is beneficial to both treatment and discovery of mechanisms of OGM in depression.

5. The disturbed interface between cognitive control and emotion processing

Major depressive disorder is characterized by a negative cognitive bias or schema, in which depressed patients are prone to negative perception bias or show difficulty in disengagement from depression-related information[74] or diminished capacity to experience pleasure (anhedonia). Although numerous studies have examined the neural basis underlying the negative cognitive bias of depression, the mechanisms remain unclear. In general, the negative cognitive biases in depression are facilitated by increased influence from subcortical emotion processing regions combined with attenuated top-down cognitive control.

5.1. Excessively emotion-processing

Previous study focused on the bottom-up emotion-processing (most notably the amygdala and fusiform gyrus) to explore the neural mechanism of depression. The hyperactivity of amygdala and fusiform gyrus is obviously correlated with the bias reflects excessive bottom-up responses to negative stimuli.

The amygdala, a brain structure which is involved in detecting emotion (for example, salience detection), interprets and maintains the emotional quality of the stimulus. Amygdala activity increases in healthy individuals during processing of emotional information. Compared to healthy controls, individuals with depression show increased activity during the perception and evaluation of, and response to negative emotion-inducing stimuli. Recent studies explored the behavioral characteristics and neural mechanism with a validated emotional face-matching task[75]. Although both MDD and HC groups were not significantly different in RT and percent correct for faces, or shapes, the MDD relative to HC subjects showed increased activity in bilateral extended amygdale during performance of a validated emotion-processing task. Besides, the MDD showed more task-related co-activation of the subgenual cingulate, which is involved in processing negative self-referential information; and less co-activation of the supragenual cingulate, which is involved in the cognitive control of emotion. Greater depressive symptom severity correlated positively with decreased FC between bilateral extended amygdala (EA) and supragenual cingulate in MDD subjects. Furthermore, the increased amygdala reactivity to negative stimuli is not only found in MDD, but also in BD (Bipolar Disorder) [76]. Other studies also indicated that amygdala hyperactivity is a neural substrate of biased attention [77, 78] and biased memory for negative stimuli[79]. Similar result was found in unmedicated remitted depressed individuals. Following sad mood induction, bilateral amygdala response during encoding of valenced words predicted increased recall of negative self-referent words for a subset of remitted depressed participants[80]. This association was not present before the sad mood induction and was not evident in individuals without a history of depression, regardless of mood state. These results suggest a role for amygdala in modulating mood-congruent memory during transient sad mood in individuals vulnerable to depression relapse.

Besides amygdala, the negative cognitive bias also linked to deficits in fusiform gyrus. Using event-related fMRI, neural responses to happy and sad facial expressions were measured in healthy individuals and individuals with major depressive disorder [78]. The study indicated that depressed but not healthy individuals demonstrated linear increases in response in right fusiform gyrus to expressions of increasing sadness. Besides, similar results was found in the neural responses of high-risk population of depression compared to low-risk group (by virtue of high and low neuroticism scores; high-N group and low-N group respectively) during the presentation of fearful and happy faces using fMRI [81]. The results indicated that the high-N group demonstrated linear increases in response in the right fusiform gyrus and left middle temporal gyrus to expressions of increasing fear, whereas the low-N group demonstrated the opposite effect. Besides, the high-N group also displayed greater responses in the right amygdala, cerebellum, left middle frontal and bilateral parietal gyri to medium levels of fearful v. happy expressions. Furthermore, the activation during negative emotional response in right fusiform gyrus reduced after antidepressant treatment.

Excessively emotion-processing models have established that the bias reflects excessive bottom-up responses to negative stimuli[74], linked to hyperactivity in the limbic regions, such as amygdala, fusiform gyrus. Individuals with depression show increased activity during the perception and evaluation of, and response to negative emotion-inducing stimuli. Therefore, they experience further negative emotion, leading to more profound depression.

5.2. Impaired cognitive control

Previous study also focused on a diminishing the top-down cognitive control (most notably prefrontal cotex) to explore the neural mechanism of depression. This attenuation in cognitive control seems to be region specific (for example, the MPFC for self-referential schemas, the DLPFC for rumination and biased processing and the VLPFC for biased attention) and curbs the top-down relationship (through the ACC and thalamus) with pertinent subcortical regions.

Compared to healthy subjects, subjects with midlife major depression showed a failure of hippocampal and anterior cingulate activation underwent positron emission tomography imaging during a control task and verbal encoding of a paragraph[82]. Another study explored relationships between amygdala and DLPFC activity during emotional and cognitive information processing in unipolar depression [83]. The results indicated that depressed subjects displayed decreased DLPFC activity on the digit-sorting task. It is consistent with previous studies[76]. Besides, one study recorded 128-channel event-related potentials while study Patients with MDD and healthy comparison subjects performed a Stroop task, modified to incorporate performance feedback[84]. The results indicated that unmedicated patients with MDD showed reduced accuracy and potentiated error-related negativity immediately after committing errors, highlighting dysfunctions in the automatic detection of unfavorable performance outcomes, and abnormal reaction to committing errors was accompanied by hyperactivation in rostral ACC and medial PFC regions 80 milliseconds after committing errors and a failure to recruit dorsolateral PFC-based cognitive control, which is consistent with previous results.

Impaired cognitive-control models have pointed out that the bias reflects impaired top-down cognitive control[85, 86], linked not only to reduced activities in cortical regions, including DLPFC, anterior cingulate cortex (ACC), and rostral ACC[84, 87, 88], but also to reduced N450 and N200 amplitudes[84]. To maintain the same level of performance as healthy subjects, individuals with major depression need increased effort to recruit more cerebral resources, suggesting that depression may impair the top-down cognitive control capacity of afflicted patients. With limited top-down cognitive control from the PFC, the consequences of maladaptive bottom-up activity persist, including enhanced amygdala reactivity (which contributes to biased attention and cognitive processing).

5.3. The disturbed interface between cognitive control and emotion processing

As noted earlier, some studies may explain the cause of depression from the perspective of abnormal emotional processing or impaired cognitive control. However, it is unclear whether emotional processing and cognitive control affect the negative cognitive bias independently or reciprocally. To answer this question, a number of researches have been carried out. Siegle et al. reported increased activity in the amygdala in response to personally relevant negative words (personal relevance rating of words) and dorsal lateral prefrontal cortex (DLPFC) hypoactivity in a cognitive control task (digit sorting)[83]. However, because the two tasks are carried out separately, it is difficult to observe how emotion processing interacts with cognitive control. Subjects were asked to respond to several targets in the context of emotional or neutral stimuli. They found that processing of emotional contexts could interfere with the processing of cognitive control; emotional interference exerted a greater influence on depressed subjects. However, Goldin et al. found that reappraisal (a strategy of cognitive control) was associated with early (0–4.5 s) prefrontal cortex responses and decreased amygdala responses, suggesting that cognitive control played an important role in mediating emotional processing[89]. Notably, by combining emotional-processing task with cognitive-control task (emotion-interference task), Fales et al. found that depressed patients showed hyperactivity in emotion-processing regions, including the amygdala, and in cognitive-control regions, including the DLPFC and dorsal anterior cingulate cortex (ACC). Their results suggest that these processes might interact with each other[89]. Although more attention was paid to activity alterations in emotion-processing and cognitive-control regions in Fales' study, the relationship between bilateral emotion processing and bilateral cognitive-control regions was somehow ignored.

Therefore, we took bilateral amygdala as regions of interest (ROIs) to determine the relationship between bilateral emotion-processing and bilateral cognitive-control regions[90]. The results indicated that depressed patients showed abnormal activities in bilateral amygdala and the right DLPFC. In addition, a significant correlation was found between the right amygdala and the right DLPFC when subjects observed the happy faces. The results suggest that the dysfunction in positive emotion-processing and cognitive-control regions may reciprocally affect negative cognitive bias. Additionally, altered positive emotional interference processing in the fronto-limbic brain circuitry might be another cause of negative cognitive bias that finally leads to depression.

To sum up, depression is characterized by a negativity bias which is a stable factor of it. The bias reflects enhanced bottom-up responses to affective stimuli, linked to deficits in amygdala and fusiform gyrus function. Alternatively, the bias also reflects impaired top-down cognitive control, linked to deficits in dorsolateral prefrontal cortex and anterior cingulate function. We recommended a new hypothesis that the occurrence of depression is caused by interaction of dysfunction in positive emotion processing brain regions and deficits in cognitive-control brain regions.

6. Future trends and conclusion

In summary, several lines of research point to separate mechanisms (for example, over generalized autobiographical memory, negative attention bias, abnormal emotion regulation, deficits in cognitive control of emotion, negative appraisal and reappraisal), in individuals with depression, that increase the salience of negative stimuli and decrease the salience of positive or rewarding stimuli. As a result, a person with depression displays a cognitive bias towards negative information and away from positive information, thus contributing to the maintenance of a depressed mood state.

In parallel with existing neurobiological models of depression that focus on affective symptomatology, Beck' s cognitive–neurobiological model suggests that cognitive biases in depression are due to maladaptive bottom-up processes that are generally perpetuated by attenuated cognitive control, and has provided an evidence-based framework to conceptualize and treat major depressive disorder. While Beck's cognitive model provides new features and benefits to understanding the symptoms and underlying neural substrates for major depressive disorder. The mechanism of various components leading to depression is still vague.

Future research should seek to identify which neurobiological mechanisms contribute to the selective processing towards negative, and away from positive, environmental stimuli.

Ackowledgements

Zhengzhi Feng was supported by the National Natural Science Foundation of China (NSFC30970898) during the empirical research. The current review was supported by the Project of Military Mental Health Research (BWS11J045).

Author details

Zhengzhi Feng[1*], Xiaoxia Wang[2], Keyu Liu[1], Xiao Liu[1], Lifei Wang[3], Xiao Chen[1] and Qin Dai[4]

*Address all correspondence to: fzz@tmmu.edu.cn

1 College of Psychology, Third Military Medical University, Chongqing, China

2 Department of Basic Psychology, College of Psychology, Third Military Medical University, Chongqing, China

3 Department of Developmental and Educational Psychology, College of Psychology, Third Military Medical University, Chongqing, China

4 Department of Nursing Psychology, School of Nursing, Third Military Medical University, Chongqing, China

References

[1] Koster, E.H.W., et al., Understanding depressive rumination from a cognitive science perspective: The impaired disengagement hypothesis. Clinical Psychology Review, 2011. 31(1): p. 138-145.

[2] Koster, E.H.W., D. R and E. Goeleven, Mood congruent attentional bias in dysphoria: Maintained attention to impaired disengagement from negative information. Emotion, 2005. 5: p. 446-455.

[3] Joormann, J. and I.H. Gotlib, Slective attention to emotional faces following recovery from depression. Journal of Abnormal psychology, 2007. 116(1): p. 80-85.

[4] AS, S., A. WY and C. S, Recognition accuracy and response bias to happy and sad facial expressions in patients with major depression. Neuropsychology, 2004. 18(2): p. 212-218.

[5] Joormann, J. and I.H. Gotlib, Emotion regulation in depression: Relation to cognitive inhibition. Cognition & Emotion, 2010. 24(2): p. 281-298.

[6] Kross, E., D. Matthews and J. Weber, Coping with emotions past: the neural bases of regulating affect associated with negative autobiographical memories. Biological Psychiatry, 2009. 65(5): p. 361-366.

[7] Habel, U., W. C and D. B, Amygdala activaton and facial expressions: Explicit emotion discrimination versus implicit emotion processing. Neuropsychologia, 2007. 45: p. 2369-2377.

[8] Gotlib, I.H. and J. Joormann, Cognition and depression: current status and future directions. Annu Rev Clin Psychol, 2010. 6: p. 285-312.

[9] Koster, E.H., et al., Mood-congruent attentional bias in dysphoria: maintained attention to and impaired disengagement from negative information. Emotion, 2005. 5: p. 446-55.

[10] Zhengzhi, F., et al., A brief introduction for attention bias theory of depression, in The Neurocognitive Basis for Depression. 2011, Science Press: Peking. p. 6-8.

[11] Jongen, E.M., et al., Attentional bias and general orienting processes in bipolar disorder. J Behav Ther Exp Psychiatry, 2007. 38: p. 168-83.

[12] Eizenman, M., et al., A naturalistic visual scanning approach to assess selective attention in major depressive disorder. Psychiatry Res, 2003. 118: p. 117-28.

[13] Zhang, H.H., J. Zhang and S. Kornblum, A parallel distributed processing model of stimulus-stimulus and stimulus-response compatibility. Cogn Psychol, 1999. 38: p. 386-432.

[14] Karparova, S.P., A. Kersting and T. Suslow, Disengagement of attention from facial emotion in unipolar depression. Psychiatry Clin Neurosci, 2005. 59: p. 723-9.

[15] Yovel, I. and M. Susan, Emotion-congruent attentional biases: the perspective of hierarchical models of emotional disorders. Personality and Individual Differences, 2005. 38: p. 785-795.

[16] Bradley, B.P., K. Mogg and S.C. Lee, Attentional biases for negative information in induced and naturally occurring dysphoria. Behav Res Ther, 1997. 35: p. 911-27.

[17] Gotlib, I.H., et al., Attentional biases for negative interpersonal stimuli in clinical depression. J Abnorm Psychol, 2004. 113: p. 121-35.

[18] Hill, A.B. and F. Dutton, Depression and selective attention to self-esteem threatening words. Personality and Individual Differences, 1989. 10: p. 915-917.

[19] Cavanagh, J. and M.W. Geisler, Mood effects on the ERP processing of emotional intensity in faces: a P3 investigation with depressed students. Int J Psychophysiol, 2006. 60: p. 27-33.

[20] MacDonald, A.W.R., et al., Dissociating the role of the dorsolateral prefrontal and anterior cingulate cortex in cognitive control. Science, 2000. 288: p. 1835-8.

[21] Rogers, M.A., et al., Executive and prefrontal dysfunction in unipolar depression: a review of neuropsychological and imaging evidence. Neurosci Res, 2004. 50: p. 1-11.

[22] Caseras, X., et al., Biases in visual orienting to negative and positive scenes in dysphoria: An eye movement study. J Abnorm Psychol, 2007. 116: p. 491-7.

[23] Rinck, M. and E.S. Becker, A comparison of attentional biases and memory biases in women with social phobia and major depression. J Abnorm Psychol, 2005. 114: p. 62-74.

[24] Dai, Q. and Z. Feng, Deficient inhibition of return for emotional faces in depression. Prog Neuropsychopharmacol Biol Psychiatry, 2009. 33: p. 921-32.

[25] Dai, Q. and Z. Feng, Dysfunctional distracter inhibition and facilitation for sad faces in depressed individuals. Psychiatry Res, 2011. 190: p. 206-11.

[26] Dai, Q., Z. Feng and E.H. Koster, Deficient distracter inhibition and enhanced facilitation for emotional stimuli in depression: an ERP study. Int J Psychophysiol, 2011. 79: p. 249-58.

[27] Dai, Q. and Z. Feng, Deficient interference inhibition for negative stimuli in depression: an event-related potential study. Clin Neurophysiol, 2011. 122: p. 52-61.

[28] Disner, S.G., et al., Neural mechanisms of the cognitive model of depression. Nat Rev Neurosci, 2011. 12(8): p. 467-77.

[29] Gross, J.J. and L. Feldman Barrett, Emotion Generation and Emotion Regulation: One or Two Depends on Your Point of View. Emotion Review, 2011. 3(1): p. 8-16.

[30] Iwabuchi, S.J., et al., Alterations in effective connectivity anchored on the insula in major depressive disorder. European Neuropsychopharmacology, 2014.

[31] Bylsma, L.M., B.H. Morris and J. Rottenberg, A meta-analysis of emotional reactivity in major depressive disorder. Clinical Psychology Review, 2008. 28(4): p. 676-691.

[32] Rottenberg, J., J.J. Gross and I.H. Gotlib, Emotion context insensitivity in major depressive disorder. Journal of Abnormal Psychology, 2005(114): p. 627-639.

[33] Cullen, K.R., et al., ABnormal amygdala resting-state functional connectivity in adolescent depression. JAMA Psychiatry, 2014.

[34] Dai, Q. and Z. Feng, More excited for negative facial expressions in depression: evidence from an event-related potential study. Clin Neurophysiol, 2012. 123(11): p. 2172-9.

[35] Joormann, J. and I.H. Gotlib, Emotion regulation in depression: Relation to cognitive inhibition. Cognition & Emotion, 2010. 24(2): p. 281-298.

[36] Murphy, F.C., A. Michael and B.J. Sahakian, Emotion modulates cognitive flexibility in patients with major depression. Psychol Med, 2012. 42(7): p. 1373-82.

[37] Seminowicz, D.A., et al., Limbic-frontal circuitry in major depression: a path modeling metanalysis. NeuroImage, 2004. 22(1): p. 409-418.

[38] Phillips, M.L., et al., Neurobiology of emotion perception II: Implications for major psychiatric disorders. Biol Psychiatry, 2003. 54(5): p. 515-28.

[39] Dai, Q. and Z. Feng, More excited for negative facial expressions in depression: evidence from an event-related potential study. Clin Neurophysiol, 2012. 123(11): p. 2172-9.

[40] Almeida, J.R.C., et al., Elevated Amygdala Activity to Sad Facial Expressions: A State Marker of Bipolar but Not Unipolar Depression. Biological Psychiatry, 2010. 67(5): p. 414-421.

[41] Wang, X., et al., Dissociable Self Effects for Emotion Regulation: A Study of Chinese Major Depressive Outpatients. BioMed Research International, 2014. 2014: p. 11.

[42] Almeida, J.R.C.D., et al., Abnormal Amygdala-Prefrontal Effective Connectivity to Happy Faces Differentiates Bipolar from Major Depression. Biological Psychiatry, 2009. 66(5): p. 451-459.

[43] Joormann, J. and W.M. Vanderlind, Emotion Regulation in Depression: The Role of Biased Cognition and Reduced Cognitive Control. Clinical Psychological Science, 2014. 2(4): p. 402-421.

[44] Aldao, A. and S. Nolen-Hoeksema, Specificity of cognitive emotion regulation strategies: A transdiagnostic examination. Behaviour Research and Therapy, 2010. 48(10): p. 974-983.

[45] Garnefski, N., et al., Cognitive emotion regulation strategies and depressive symptoms: differences between males and females. Personality and Individual Differences, 2004. 36(2): p. 267-276.

[46] Ehring, T., et al., Characteristics of emotion regulation in recovered depressed versus never depressed individuals. Personality and Individual Differences, 2008. 44(7): p. 1574-1584.

[47] Liverant, G.I., et al., Emotion regulation in unipolar depression: The effects of acceptance and suppression of subjective emotional experience on the intensity and duration of sadness and negative affect. Behaviour Research and Therapy, 2008. 46(11): p. 1201-1209.

[48] Beblo, T., et al., Increased suppression of negative and positive emotions in major depression. J Affect Disord, 2012. 141(2-3): p. 474-9.

[49] Kwon, H., et al., Cultural and gender differences in emotion regulation: relation to depression. Cogn Emot, 2013. 27(5): p. 769-82.

[50] Kwon, H., et al., Cultural and gender differences in emotion regulation: relation to depression. Cogn Emot, 2013. 27(5): p. 769-82.

[51] Werner-Seidler, A., et al., An investigation of the relationship between positive affect regulation and depression. Behav Res Ther, 2013. 51(1): p. 46-56.

[52] Beblo, T., et al., Increased suppression of negative and positive emotions in major depression. J Affect Disord, 2012. 141(2-3): p. 474-9.

[53] Ehring, T., et al., Emotion regulation and vulnerability to depression: spontaneous versus instructed use of emotion suppression and reappraisal. Emotion, 2010. 10(4): p. 563-72.

[54] Abler, B., et al., Habitual emotion regulation strategies and depressive symptoms in healthy subjects predict fMRI brain activation patterns related to major depression. Psychiatry Res, 2010. 183(2): p. 105-13.

[55] Hopp, H., A.S. Troy and I.B. Mauss, The unconscious pursuit of emotion regulation: Implications for psychological health. Cogn Emot, 2011. 25(3): p. 532-45.

[56] Dennis, T.A., Interactions between emotion regulation strategies and affective style: Implications for trait anxiety versus depressed mood. Motivation and Emotion, 2007. 31(200-207).

[57] Hervas, G. and C. Vazquez, Low spirits keep rewards subdued: decreases in sensitivity to reward and vulnerability to dysphoria. Behav Ther, 2013. 44(1): p. 62-74.

[58] Ochsner, K.N., J.A. Silvers and J.T. Buhle, Functional imaging studies of emotion regulation: a synthetic review and evolving model of the cognitive control of emotion. Ann N Y Acad Sci, 2012. 1251: p. E1-24.

[59] Pinto-Meza, A., et al., Evaluation of behavioural activation and inhibition systems by telephone. Revista de Psiquiatría y Salud Mental (English Edition), 2009. 2(2): p. 66-71.

[60] Xiaoxia, W., et al., Abnormal resting state of MDD: the correlation with behaviral inhibition/ activation. Journal of Third Military Medical University, 2014.

[61] Linden, D., EPA-1819-Regulation of emotion networks in depression. European Psychiatry, 2014. 29, Supplement 1: p. 1.

[62] Conway, M.A. and C.W. Pleydell-Pearce, The construction of autobiographical memories in the self-memory system. Psychol Rev, 2000. 107(2): p. 261-88.

[63] Williams, J.M.G., et al., Reduced autobiographical memory specificity and affect regulation. Cognition and Emotion, 2006. 20(3-4): p. 402-429.

[64] Williams, J.M., et al., Autobiographical memory specificity and emotional disorder. Psychol Bull, 2007. 133(1): p. 122-48.

[65] Sumner, J.A., J.W. Griffith and S. Mineka, Overgeneral autobiographical memory as a predictor of the course of depression: a meta-analysis. Behav Res Ther, 2010. 48(7): p. 614-25.

[66] Raes, F., et al., A sentence completion procedure as an alternative to the Autobiographical Memory Test for assessing overgeneral memory in non-clinical populations. Memory, 2007. 15(5): p. 495-507.

[67] Wang, F., et al., A study of the paradigm for assessing overgeneral autobiographical memory in Chinese depressive mood populations. Chin J Behav Med & Brain Sci, 2010. 3(19): p. 245-247.

[68] Tulving, E., Elements of Episodic Memory. 1985: OUP Oxford.

[69] Cabeza, R. and J.P. St, Functional neuroimaging of autobiographical memory. Trends Cogn Sci, 2007. 11(5): p. 219-27.

[70] St, J.P., P.A. Kragel and D.C. Rubin, Dynamic neural networks supporting memory retrieval. Neuroimage, 2011. 57(2): p. 608-16.

[71] Lemogne, C., et al., Medial prefrontal cortex and the self in major depression. J Affect Disord, 2012. 136(1-2): p. e1-e11.

[72] Dalgleish, T., et al., Reduced specificity of autobiographical memory and depression: the role of executive control. J Exp Psychol Gen, 2007. 136(1): p. 23-42.

[73] Whalley, M.G., M.D. Rugg and C.R. Brewin, Autobiographical memory in depression: an fMRI study. Psychiatry Res, 2012. 201(2): p. 98-106.

[74] Disner, S.G., et al., Neural mechanisms of the cognitive model of depression. Nat Rev Neurosci, 2011. 12(8): p. 467-77.

[75] Matthews, S.C., et al., Decreased functional coupling of the amygdala and supragenual cingulate is related to increased depression in unmedicated individuals with current major depressive disorder. Journal of Affective Disorders, 2008. 111(1): p. 13-20.

[76] Savitz, J. and W.C. Drevets, Bipolar and major depressive disorder: neuroimaging the developmental-degenerative divide. Neurosci Biobehav Rev, 2009. 33(5): p. 699-771.

[77] Dannlowski, U., et al., Amygdala reactivity to masked negative faces is associated with automatic judgmental bias in major depression: a 3 T fMR1 study. Journal of Psychiatry & Neuroscience, 2007. 32(6): p. 423-429.

[78] Surguladze, S., et al., A differential pattern of neural response toward sad versus happy facial expressions in major depressive disorder. Biological Psychiatry, 2005. 57(3): p. 201-209.

[79] Hamilton, J.P. and I.H. Gotlib, Neural substrates of increased memory sensitivity for negative stimuli in major depression. Biological Psychiatry, 2008. 63(12): p. 1155-1162.

[80] Ramel, W., et al., Amygdala reactivity and mood-congruent memory in individuals at risk for depressive relapse. Biological Psychiatry, 2007. 61(2): p. 231-239.

[81] Yang, T.T., et al., Increased amygdala activation is related to heart rate during emotion processing in adolescent subjects. Neuroscience Letters, 2007. 428(2-3): p. 109-114.

[82] Bremner, J.D., et al., Deficits in hippocampal and anterior cingulate functioning during verbal declarative memory encoding in midlife major depression. American Journal of Psychiatry, 2004. 161(4): p. 637-645.

[83] Siegle, G.J., et al., Increased amygdala and decreased dorsolateral prefrontal BOLD responses in unipolar depression: Related and independent features. Biological Psychiatry, 2007. 61(2): p. 198-209.

[84] Holmes, A.J. and D.A. Pizzagalli, Spatiotemporal dynamics of error processing dysfunctions in major depressive disorder. Archives of General Psychiatry, 2008. 65(2): p. 179-188.

[85] Clark, D.A. and A.T. Beck, Cognitive theory and therapy of anxiety and depression: Convergence with neurobiological findings. Trends in Cognitive Sciences, 2010. 14(9): p. 418-424.

[86] Marazziti, D., et al., Cognitive impairment in major depression. European Journal of Pharmacology, 2010. 626(1): p. 83-86.

[87] Beevers, C.G., et al., Depression Symptoms and Cognitive Control of Emotion Cues: A Functional Magnetic Resonance Imaging Study. Neuroscience, 2010. 167(1): p. 97-103.

[88] Lee, B.T., et al., Neural correlates of affective processing in response to sad and angry facial stimuli in patients with major depressive disorder. Progress in Neuro-Psychopharmacology & Biological Psychiatry, 2008. 32(3): p. 778-785.

[89] Goldin, P.R., et al., The neural bases of emotion regulation: Reappraisal and suppression of negative emotion. Biological Psychiatry, 2008. 63(6): p. 577-586.

[90] Liao, C., et al., Dysfunction of Fronto-Limbic Brain Circuitry in Depression. Neuroscience, 2012. 201: p. 231-238.

Chapter 6

Impaired Mental Processing Speed With Moderate to Severe Symptoms of Depression

Tabitha W. Payne and Madeline Thompson

Additional information is available at the end of the chapter

http://dx.doi.org/10.5772/59597

1. Introduction

One way to better understand clinical disorders is to combine research from cognitive experiments to dissociate information processing impairments for various conditions, such as major depressive disorder. Ruminations, which are associated with depression, place an increased load on basic cognitive processes, such as working memory and attention, which are necessary for more advanced reasoning and problem solving. Research has revealed a connection between performance decrements in an array of activities requiring effortful processing, or processes that rely on working memory and controlled attention [1]. Depression impairs such cognitive functions as verbal fluency [2,3], verbal memory [4], mental flexibility [5], effortful processes and executive functions [6].

While working memory and attention are important contributing processes to fluid intelligence [7], mental processing speed also contributes to variance in higher level reasoning and problem solving [8,9], yet little research has been devoted to understanding mental speed in the depressed individual. Measures of mental processing speed can include reaction time for decision making. For example, in a study by Kalb, Dorner, and Kalb (2006), depressed patients were administered both simple and choice reaction time measures [10]. Depressed patients were slower than control participants on measures using briefly presented stimuli. After administration of antidepressants, the depressed patients showed decrease in reaction time, linking the medication to better speed of processing, however, performance errors increased. The results of this research indicate differences in information processing occurring very early at the stage of pre-attentive processing.

Another way to examine mental speed and depression is to use inspection time measures [11]. Inspection time assessments entail manipulating the exposure duration of stimuli. Most inspection time measures require a decision, such as auditory or visual discrimination of masked stimuli. Inspection time is considered to be an indirect measure of neural firing speed in the brain, specifically for the processes necessary for an accurate decision. Individuals with fast neural speed can make an accurate representation of the stimulus in less time, which could facilitate subsequent decision making and problem solving. If depression is characterized by slowed mental speed on inspection time measures, then decision making is also impaired. Depressive symptoms, as characterized by the Diagnostic and Statistical Manual of Mental Disorders, Fifth Edition (DSM-V), include feeling slowed down, fatigued, and inhibited concentration and thinking [12]. According to **cognitive slowing theory** [13], depression is linked to neural slowing in the brain, as opposed to the **dysfunctional basal ganglia theory** proposed by Lohr and colleagues (2013), in which sluggishness in depression is due to motor impairment, similar to Parkinson's Disease in which mental speed is not impaired, only movement [14]. Evidence for the later theory comes from a sample of patients with depression that performed simple reaction time tasks that involved detection of a visual target. Patients performed similar to controls on reaction time measures, however, still performed more poorly on a number of movement measures [14]. A criticism of this research is the lack of variation in mental speed measures used. It is possible that performance deficits are specific to the speeded decision, for example detection versus discrimination judgments. Mental speed differences could be associated with depression, but researchers need to examine a more expansive set of mental speed paradigms.

Evidence for cognitive slowing has been found in a study by Tsourtos, Thompson, and Stough (2002), in which inpatients with depressive disorder were compared with controls on visual inspection time tasks using various sizes of lines [13]. Here, the depressed patients showed decrements on mental speed tasks with exposure durations as short as 60 ms. The visual inspection time measures required discrimination judgments for briefly presented lines.

The focus of the current research endeavor was to examine the potential relationship between depression and mental speed using inspection time measures, and expanding the variation to speeded detection, identification, and discrimination of letters presented at durations of 80 to 16 ms. Since inspection time for line discrimination was found to be impaired in depressed patients [13], letter discrimination is also expected to be impaired in the current study. Since the other measures, detection and identification, are unexplored, it is unknown whether they will be impacted by depression. Unlike the existing research examining inspection time, this study used a college student population, and self-report on the Beck Depression Inventory [15]. It is important to examine the potential impairments in a non-clinical sample, especially students, who must rely on peak cognition to maintain academics in a college setting.

2. Methods

2.1. Participants

The sample consisted of 217 college student participants from a highly selective liberal arts college in the Midwest. Age ranged 18-23, with a gender distribution of 144 females, 73 males.

Participants signed up online or called for individual appointments, and were given the option of receiving monetary compensation or research credit in a psychology course, if enrolled.

2.2. Measures

Beck Depression Inventory. After a detailed demographic survey with medical history, participants were given the computerized version of the mood inventory in a private setting and were encouraged to work at their own pace. The BDI asks patients to respond to 21 scales from 0-3. As an example item on the BDI, participants note which correspond to their life with (0) I do not feel sad; (1) I feel sad; (2) I am sad all the time and I can't snap out of it; (4) I am so sad and unhappy that I can't stand it. Scores range from 0-63, with 14-19 qualifying for mild depression, 20-28 for moderate depression, and 29 and up for severe depression. Reliability of the BDI is quite high, with a test-retest correlation of 0.93, and it also correlates with other depression measures, indicating a reliable concurrent validity [15].

Snellen Test for Visual Acuity Screening. After informed consent each participant was administered a visual acuity test using a standardized Snellen eye chart to identify individuals with visual deficits, with a criterion of 20/20 vision required for inclusion in the mental speed measures. Participants read letters of varying sizes (line by line). All participants were informed prior to the session to wear corrective eyewear for this assessment.

Visual Inspection Time for Letter Detection and Identification. This computerized assessment measures two variables: accuracy for detection of briefly presented letters and accuracy for the verbal identification of the letters present. Additionally this measure required participants to identify letters appearing in the center of the computer screen for varying amounts of exposure or "inspection" time, with the durations decreasing in trial blocks. There were a total of 5 blocks, with each including 15 trials. For each block, 5 trials were "blank" trials in which no letter appeared, while 10 trials contained a target letter. Each target letter was presented twice within a block of trials, with the target letters including X, Z, H, K, and E, in size 18 font. Prior to actual test trials the participants were provided with practice trials to demonstrate, with a 500 ms inspection time. After practice, inspection time for the target letters was 80, 50, 30, 20, and 16 ms, with blocks presenting inspection times for decreasing amounts as the task continued.

For each trial a "Ready" screen appeared with a prompt for participants to self-initiate the trial sequence. Once a trial was initiated there was a refractory period of 500ms. Next a forward visual mask, a"#" sign, was displayed for 30 ms, followed by a potential letter presentation or a blank screen for either 80, 50, 30, 20, or 16 ms, depending on the block. A backward visual mask followed, which was another "#" sign for 300ms. The general perception in this task is you must decide if a letter is present between the "#". Presence of blank trials was random for each specific inspection time block. Responses for letter detection were indicated by a key press on a computer key pad, with designated keys marked ("1" for yes or "3" for no). If participants indicated a letter was detected, the next instruction was to attempt to identify the letter by choosing it on the keyboard. Feedback was provided on accuracy at the end of each trial. Measures attained by this task are accuracy for detection and identification, for each presentation duration block (80-16 ms). Refer to Figure 1 for a representation of stimulus events.

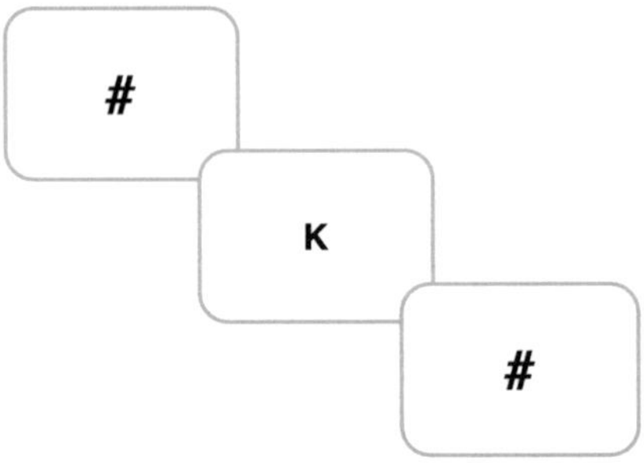

Figure 1. Inspection Time Event Sequence. The boxes represent the series of stimulus events that appear on the computer screen in the mental speed assessment for letter detection and identification. The target letter stimulus is embedded between the presentation of a forward and backward mask (#).

Visual Inspection Time for Letter Discrimination. A second computerized task for visual inspection time was given to assess speeded discrimination. This assessment required participants to decide whether two briefly presented letters pairs are comprised of same or different letters. Similar to the detection/identification tasks, blocks included 15 trials, with descending inspection time durations of 80, 50, 30, 20, and 16 ms. The target letter pairs were XX, KX, EH, ZZ, or XK presented in size 18 font. Also, a practice block was provided in which the target letter pair appeared for 500 ms. Trial parameters were similar to those in the initial detection version, however, with the larger area of two letters, the mask was designed larger (36 font) in order to mask the stimuli. Participants indicated a decision by pressing either "1" for same or "3" for different.

2.3. Procedure

Participants were tested in individual appointments at the Kenyon College Cognition Lab, in sound attenuated rooms, and were instructed to work at a comfortable pace. After completing surveys, participants were screened on the Snellen Eye Chart. The participants then completed the first computerized inspection time task for detection and identification. For the performance measures, each participant was seated in a stationary chair 17 inches from the monitor to control for visual angle. The participants then completed the first computerized inspection time task for detection and identification, which lasted approximately 10 to 15 minutes. Then after a break, a second computerized task was administered for discrimination. At debriefing, participants were provided with contact information for counseling services in the local area and encouraged to seek support if concerned about survey responses.

3. Results

3.1. Beck depression inventory

Scores on the BDI had a mean of 9.67, with a standard deviation of 7.87, a range of 0-39, and high internal consistency with Chronbach's Alpha of 0.895. BDI scores were positively skewed, with 92 participants qualifying for minimal depression, 97 for mild depression, 27 for moderate, and 5 for severe (see Figure 2 for participant frequencies on BDI categories). With so few participants meeting criterion for severe depression, all subsequent analyses for hypothesis testing compared those in both the moderate and severe group, giving a total of 32 in the depressed group to compare with those not qualifying at that intense of a criterion level on the BDI. For this sample, 14.75% fall into the category of moderate to severe depression.

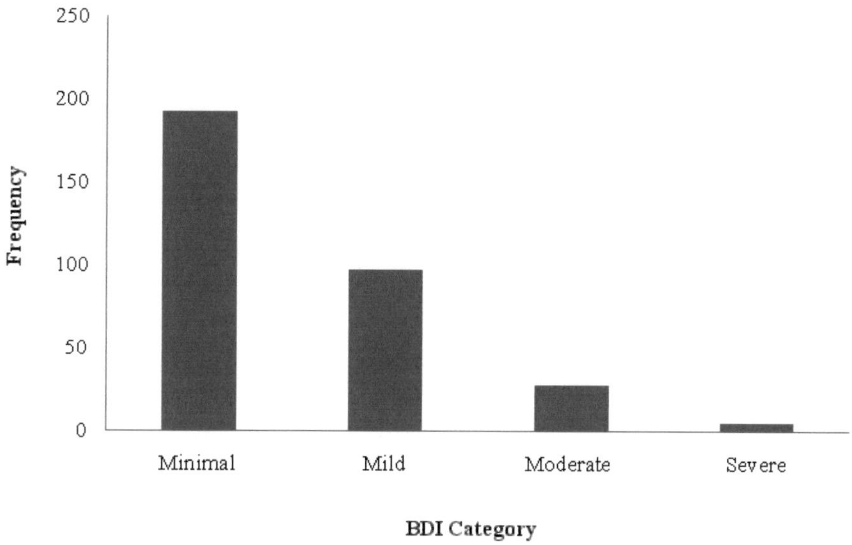

Figure 2. Beck Depression Inventory Scores. Frequency of participants meeting criterion for each identified group on the BDI assessment.

3.2. Inspection time measures

Letter Detection, Identification, and Discrimination Speed. Total performance for all participants for each of the 3 inspection time measures is shown in Figure 3 (means and standard error as a function of stimulus exposure duration). A General Linear Model ANOVA with repeated measures, with a 3 x 5 within-subject design was used to examine how performance accuracy on the 3 speed measures varies with inspection time (80, 50, 30, 20, and 16 ms). There was a significant main effect of task type ($F(2,209)=475.80$, $p<.001$), as well as a significant main effect of inspection time duration ($F(4,209)=100.95$, $p<.001$). A significant interaction was present between task type and inspection time duration ($F(8,209)=32.50$, $p<.001$). Refer to Figure 3 for data on all 3 speed measures. Note that the detection version has much higher accuracy overall, with identification and discrimination having lower accuracy. All tasks were significantly different from one another. Exposure time led to decreased accuracy as it got shorter, and there seems to be a more drastic drop in identification on the 16 ms condition than the discrimination task. In summary, these findings are consistent with the notion that the 3 inspection time measures are unique in that they measure different cognitive processes, and each has a unique relationship with exposure duration. Since the more difficult processing would occur in the identification and discrimination versions, perhaps group differences are more likely to appear in those tasks.

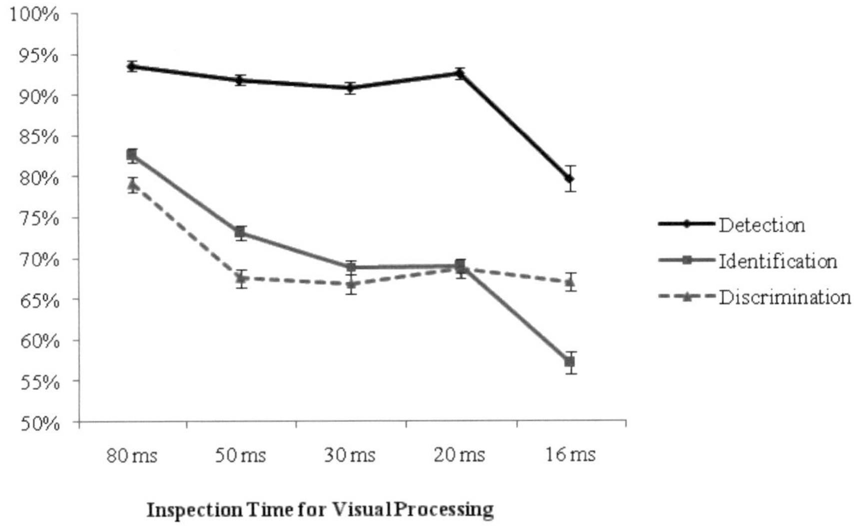

Figure 3. Inspection Time Measures. Mean accuracy with standard error bars for each inspection time task as a function of stimulus exposure duration. Note the near ceiling performance on speeded letter detection, while the more complex speed measures have lower accuracy.

3.3. Group performance data

Overall Inspection Time Accuracy. The central hypothesis was that mental speed, as measured by performance accuracy on inspection time tasks, would be different for those meeting criterion for severe depression on the BDI than those who did not meet criterion. Specifically, those with depression were predicted to perform worse, at least for the discrimination speed measure. A 3 x 2 General Linear Model ANOVA with repeated measures design was used with 3 levels of task type (detection, identification, and discrimination) and 2 participant groups (no diagnosis and moderate to severe depression). Refer to Figure 4 for mean accuracy by group. A significant effect of test type was found, ($F(2,203)=261.76$, $p<.001$). Although there was no significant main effect of participant group found on overall speed, ($F(1,203)=1.58$, $p=.21$), there was a significant interaction between participant group status and test type, ($F(2,203)=3.93$, $p=.02$), with post hoc analysis revealing a significant group difference for the discrimination version only, with the depressed group performing worse ($p<.05$). More specific analyses were conducted to examine group performance on each speed measure as a function of inspection time below.

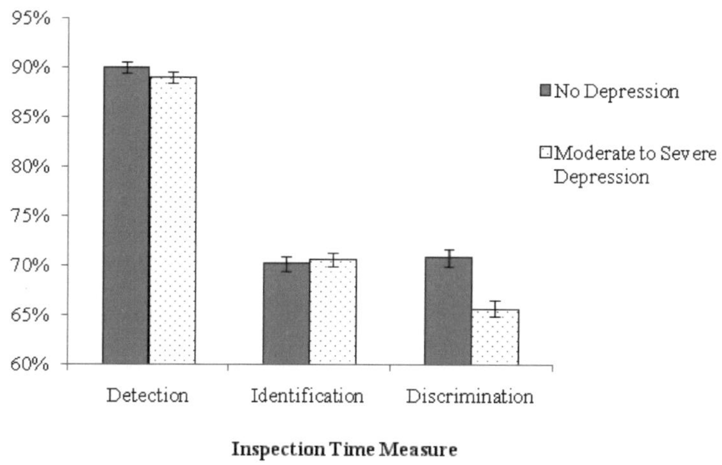

Figure 4. Group Differences in Overall Inspection Time. Mean accuracy for 2 participant groups is depicted for all 3 visual inspection time measures with standard error bars. Note the difference in overall accuracy between the 3 tasks, with accuracy being significantly lower on the identification and discrimination versions. Most notably, the participants meeting criterion for severe depression on the BDI score significantly lower on the speeded discrimination task in comparison to the non-depressed group.

Inspection Time for Letter Detection. Additional analyses were performed to examine specific group data patterns across the 5 inspection time durations (80-16 ms) for each of the mental speed assessments. A 2 x 5 General Linear Model Anova with repeated measures was used to examine performance accuracy for all 2 participant types, (no diagnosis and moderate to severe depression), as a function of inspection time intervals. The analysis resulted in no significant main effect of diagnostic criterion on overall detection accuracy, ($F(1,205)=.41$, $p=.21$). A main effect was found for inspection time duration, ($F(4.205)=24.62$, $p<.001$), with accuracy dropping for all participants with shorter exposure durations. There was no interaction between participant status and inspection time durations, ($F(4,205)=.13$, $p=.97$). Refer to Figure 5 for means accuracy for groups across inspection times for the speeded letter detection version.

Inspection Time for Letter Identification. Again, a 2 x 5 (group diagnostic status by inspection time duration) General Linear Model ANOVA with repeated measures design resulted in no significant main effect of group on accuracy overall, ($F(1,205)=.053$, $p=.82$), however there was a main effect of stimulus exposure duration, ($F(4,205)=74.11$, $p<.001$), indicating that participant accuracy decreased as the stimulus duration decreased. No significant interaction was found, ($F(4,205)=1.46$, $p=.21$) between group and duration on accuracy for identification of letters. Figure 6 shows the highly similar performance pattern between the 2 groups on speeded letter identification. Although the depressed group is actually better at 80 ms at the beginning of the task, no performance differences occur afterward.

Inspection Time for Letter Discrimination. A final 2 x 5 (group by exposure time) General Linear Model ANOVA with repeated measures did result in a significant main effect of diagnostic group with the depressed group performing worse, ($F(1, 203)=5.13$, $p=.025$), as well

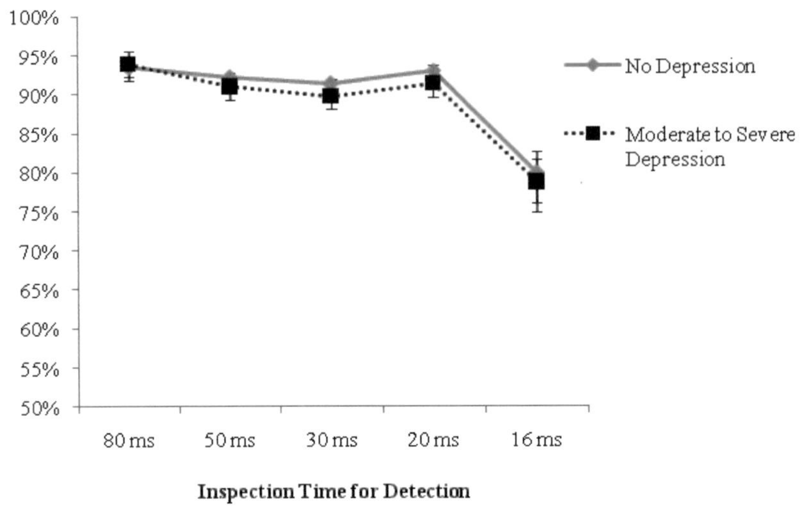

Figure 5. Inspection Time for Letter *Detection*. Mean accuracy for letter detection as a function of exposure duration is shown with standard error bars.

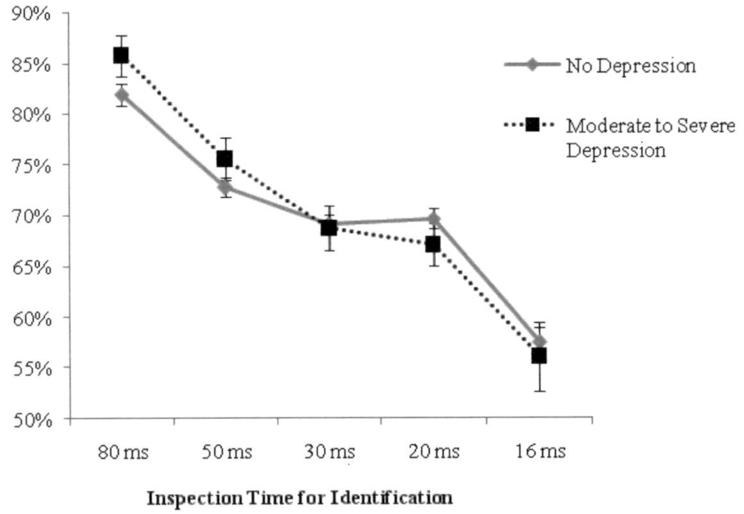

Figure 6. Inspection Time for Letter *Identification*. Mean accuracy for letter identification as a function of inspection time is presented with standard error bars.

as a main effect of exposure duration (F(4,203)=22.16, p <.001), indicating that participant accuracy decreased as the stimulus duration decreased. Similar to the data pattern with the other inspection time measures, there was no significant interaction between group and stimulus duration, (F(4,203)=.43, p=.79). Refer to Figure 7 for speeded letter discrimination accuracy by participant group and inspection time. Post hoc analyses reveal the two groups are significantly different on all conditions of the discrimination version except the initial 80 ms condition (p <.05).

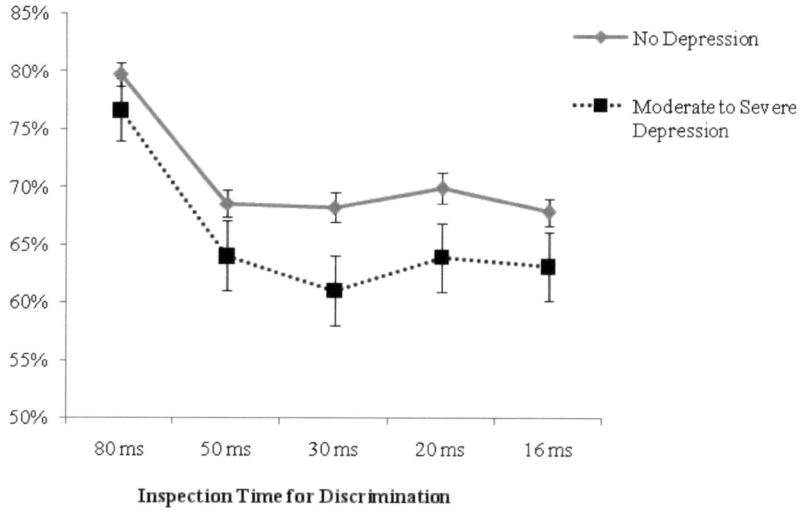

Figure 7. Inspection Time for Letter *Discrimination*. Mean accuracy is presented, along with standard error bars for accuracy at discriminating simultaneously presented letter pairs. Depressed participants are lower in accuracy in all conditions with inspection times shorter than 80 ms.

4. Discussion

Results of this study confirm that individuals reporting symptoms of moderate to severe depression, based on the Beck Depression Inventory criterion, do have impaired mental speed, as measured by the discrimination version of the inspection time measure. By including the additional speed measures for detection and identification, the goal was to see if depression would also be associated with changes in the more basic speed measures.

This finding is consistent with previous research using patient samples [13]. Speeded discrimination is a standard paradigm for inspection time measurement, is widely used, and has been found to be sensitive to participant variables, such as age and other diagnostic criteria, and has now been found to be sensitive to depression symptoms.

With multiple studies now concluding that depressive disorder is associated with impairments in mental processing speed, it is important to focus on why the relationship exists. Based on both the description in the DSM-V, and reported symptoms of patients and volunteer participants, the experience of depression is that of feeling slowed, lethargic, and fatigued. In this sense, depression may be categorized by a general nervous system slowing in the brain, which impacts neural firing necessary for processing briefly presented stimuli, (as measured by the inspection time tasks). A depressed individual needs more time with the stimuli in order to make an accurate decision about features. It may possibly take more time to form an accurate, detailed representation of stimuli when depression is more severe. The findings of this study are consistent with the **cognitive slowing hypothesis** [13], in which mental speed is disrupted, and in turn, may lead to psychomotor slowing. Another possibility is that depression, as a ruminative disorder, like anxiety, leads to detriments in working memory and attention, which in turn, leads to difficulty on any cognitive assessment that requires sustained or heightened focus. Since we know that depression and anxiety are both ruminative conditions, it would be beneficial to examine how these two disorders are related to mental speed. No research has yet been conducted on anxiety (state or trait) and mental speed. Dissociation in group performance, based on either depression or anxiety disorder would indicate that the two mood disorders may be unique in associated cognitive impairments, while similar performance decrements would indicate that the rumination hypothesis, which puts load on working memory and focus of attention, might be why depressed individual are showing impaired discrimination speed. The challenge in conducting such a study would require an adequate sample size to compare mental processing speed for depressed, anxious, and perhaps a comorbid depression-anxiety group, along with a comparison group that has no mood diagnosis. High rates of comorbidity exist between anxiety and depression [16], and studies place the rate of individuals diagnosed with both depressive and anxiety disorders at over 50 percent [17]. Individuals with anxiety and depression comorbidity experience more severe levels of symptoms associated with both conditions [18]. Comorbidity results in not only greater severity of symptoms, but also less effective treatments and treatment outcomes [19] and greater impairment of functioning [20]. The neural effects of the comorbidity of the two disorders have gone relatively unstudied, and gaps exist in literature addressing cognitive functioning in individuals with both anxiety and depression symptoms.

What some may identify as a central limitation of this study, the student sample, may also be considered a strength. The participants had to meet strict academic criterion to enroll in a selective private liberal arts college, and should score high in intelligence and reasoning. Even though the sample is not comprised of patients in clinical treatment institutions, differences in performance are consistent with previous findings [13], and it is likely that depression symptoms and cognitive deficits should be exacerbated in a more representative community or patient sample. Even with a seemingly healthy, capable subject pool, moderate to severe symptoms are associated with reduced mental speed performance. We assume that the differences found in discrimination speed would have a larger effect size with a sample more representative of the general public. Since there were only 5 individuals meeting criterion for severe depression, the fact that moderate depression symptoms yield impaired discrimination speed performance is an indication that one does not need to be an extreme case or patient in a clinic in order for cognition to be impaired. It is also important to note that with a patient sample, clinicians may be more able to assess the length of the depressive episodes and possibly gather more information as to whether depressed symptoms are a first time experience or related to isolated events, and thus attributed to state depression as opposed to trait depression. Knowing if the depression is a first time episode may also be critical to determine because research has shown that the first incidence may not impair cognition on a visual search paradigm [21]. The value of examining a student population is that the sample requires good cognition for everyday problem solving related to the goal of achieving a diploma. Therefore, any problems that are observed with basic cognitive processes are notable, as they may be associated with difficulty in higher-order decision making and reasoning necessary to complete assignments for courses.

Although participants provided only self-reported symptoms for depression, the findings are consistent with other research using undergraduate samples and self-reported data on the BDI finding cognitive impairments [22], as well as research using patient populations that have been assessed for depressive disorder via a structure diagnostic interview from a clinician [13]. Results of a study comparing medicated and non-medicated patients with depression indicate that both show impairments in relation to non-patient controls, however, those who were medicated performed better than those not [13]. Future research should examine potential changes in performance before and after medication as a treatment to better understand the potential for intra-individual improvements in mental speed. Findings of the current research with student samples indicate that this population is also vulnerable to depression and cognitive impairment, and thus administrators and educators should be aware that young adult college students also suffer, and that the impact may be seen in academics.

Author details

Tabitha W. Payne* and Madeline Thompson

*Address all correspondence to: paynet@kenyon.edu

Kenyon College Cognition Lab, Depts of Psychology & Neuroscience, Gambier, Ohio, USA

References

[1] Gallassi R, Morreale A, Pagni P. The relationship between depression and cognition. Arch Gerontol Geriatr Suppl. 2001;7:163-171.

[2] Fossati P, Guillaume IB, Ergis AM, Allilaire JF. Qualitative analysis of verbal fluency in depression. Psychiatry Res. 2003;117(1):17-24.

[3] Videbech P, Ravnkilde B, Kristensen S, et al. The Danish PET/depression project: poor verbal fluency performance despite normal prefrontal activation in patients with major depression. Psychiatry Res. 2003;123(1):49-63.

[4] Hermens DF, Naismith SL, Redoblado Hodge MA, Scott EM, Hickie IB. Impaired verbal memory in young adults with unipolar and bipolar depression. Early Interv Psychiatry. 2010;4(3):227-233.

[5] Airaksinen E, Larsson M, Lundberg I, Forsell Y. Cognitive functions in depressive disorders: evidence from a population-based study. Psychol Med. 2004;34(1):83-91.

[6] McIntyre RS, Cha DS, Soczynska JK, et al. Cognitive deficits and functional outcomes in major depressive disorder: determinants, substrates, and treatment interventions. Depress Anxiety. 2013;30(6):515-527.

[7] Kane MJ, Hambrick DZ, Tuholski SW, Wilhelm O, Payne TW, Engle RW. The generality of working memory capacity: a latent-variable approach to verbal and visuospatial memory span and reasoning. J Exp Psychol Gen. 2004;133(2):189-217.

[8] Payne TW, Smith G. Inspection time for verbal stimuli: Letter detection, identification, and discrimination speed. Nova Science Publishers: NY: In L.T. Klein & V. Amato (Eds.), Language Processing: New Research; 2013:87-104.

[9] Sheppard LD, Vernon PA. Intelligence and speed of information-processing: A review of 50 years of research. Personality and Individual Differences. 2008;44(3): 535-551.

[10] Kalb R, Dorner M, Kalb S. Opposite effects of depression and antidepressants on processing speed and error rate. Prog Neuropsychopharmacol Biol Psychiatry. 2006;30(2):244-250.

[11] O'Connor TA, Burns NR. Inspection time and general speed of processing. Personality and Individual Differences. 2003;35(3):713-724.

[12] American Psychiatric Association. Diagnostic and Statistical Manual of Mental Disorders (5th ed.). Arlington, VA: American Psychiatric Publishing; 2013.

[13] Tsourtos G, Thompson JC, Stough C. Evidence of an early information processing speed deficit in unipolar major depression. Psychol Med. 2002;32(2):259-265.

[14] Lohr JB, May T, Caligiuri MP. Quantitative assessment of motor abnormalities in untreated patients with major depressive disorder. J Affect Disord. 2013;146(1):84-90.

[15] Beck AT, Ward CH, Mendelson M, Mock J, Erbaugh J. An inventory for measuring depression. Arch Gen Psychiatry. 1961;4:561-571.

[16] Kessler RC, Chiu WT, Demler O, Merikangas KR, Walters EE. Prevalence, severity, and comorbidity of 12-month DSM-IV disorders in the National Comorbidity Survey Replication. Arch Gen Psychiatry. 2005;62(6):617-627.

[17] Mineka S, Watson D, Clark LA. Comorbidity of anxiety and unipolar mood disorders. Annu Rev Psychol. 1998;49:377-412.

[18] Kleiman EM, Riskind JH. Cognitive vulnerability to comorbidity: looming cognitive style and depressive cognitive style as synergistic predictors of anxiety and depression symptoms. J Behav Ther Exp Psychiatry. 2012;43(4):1109-1114.

[19] Lydiard RB, Brawman-Mintzer O. Anxious depression. J Clin Psychiatry. 1998;59 Suppl 18:10-17.

[20] Hofmeijer-Sevink MK, van Oppen P, van Megen HJ, et al. Clinical relevance of comorbidity in obsessive compulsive disorder: the Netherlands OCD Association study. J Affect Disord. 2013;150(3):847-854.

[21] Hammar A, Kildal AB, Schmid, M. Information processing in patients with first episode major depression. Scandinavian J Psychology. 2012;53: 445-449.

[22] Sullivan B, Payne TW. Affective disorders and cognitive failures: a comparison of seasonal and nonseasonal depression. Am J Psychiatry. 2007;164(11):1663-1667.